深圳市海阅通文化传播有限公司 / 编著

REDEFINITION
再定义 奢华宅邸
OF LUXURIOUS MANSION

華中科技大學出版社
http://www.hustp.com

ENJOY LIFE, ENJOY DESIGN

Everyone expects life could be warm and peaceful, although world keep bringing us confusion and reluctance. We need creativity and wisdom, which bloom the design and make our life develop toward the ideal world in each second.

Currently, Chinese interior design is at its turning point from the dominant and intrinsic view. From dominant view, the interior design has turned its main line from visual to cultural development, no longer blindly pursues the luxury and materialism, as well as the foreign trends, gradually forming its own style. From intrinsic view, the interior design has walked out of its blindness and into its own cultural exploring. Though our pursue still aren't very firm, the exploration is still the first step to breakthrough, and the cultural and design style will become a system tailored to our situation, as a source of unique design concept in our country.

While we are exploring the path of development, all kind of ideas will be a footstone to reach the ideal world. Now, it's time for Chinese designers to go out of the chaos. In the Decorative Art Trend Conference of 2012 Guangzhou International Design week, the Chinese top designers representives Qiu Deguang, Steve Leung and Wu Bin showed us their excellet works. Joint with FENDI, VERSACE, ARMANI and other hi-end fashion brand in the household design, Steve Leung blended his works with the essence of each brand. Since the luxrious house is no longer a wealth flaunt of the owner, it will be a significant existence with great texture. Through the material choice, the color allocation, the application of line and light, the fusion of different furniture, artworks and the emphasis on details and quality, design will reveal owner's life attitude while satisfying the owner's vision, listening and smelling, leting him experience a home full of emotions. All these above express the redefinition of luxury. Every improvement is the start of the next stage till it leads us to a new style of design. The designers fit their life understandings in works, so that their creativities could offer the furious dynamic and warmth to space. Designing a humane and luxrious space is just the process of enjoying life and enjoying design!

Steve Leung often said "enjoy life, enjoy design", which is explained like "design comes from life, and only with understandings of life, one can catch the essence of design. In return, design can lead life, the real excellent design can create more possibilities for people to enjoy life." We aren't the dominators of nature, but we can decide our life. Wish more excellent designs could bring us the happier life.

REDEFINITION OF L

享受生活 享受设计

我们都渴望生活能犹如期望的那样温馨与恬静，只是世界能提供给我们的更多的是混乱与无奈。我们需要创造，需要智慧，而这些促成了我们的设计，它让我们的生活每一刻都在向我们理想的世界发展。

中国当前的室内设计处在一个转型期。该转型表现在显性与隐性两个方面。从显性层面来看，室内设计从以外观设计为主线转型为以文化设计为主线。设计风格不再一味地追求堆砌出来的奢华，也不再盲目地追求外来的设计流派，而是日趋形成自己的风格。从隐性层面来看，室内设计从文化相互碰撞的盲目期转向以某种文化为主线的探索期。虽然目前我们对文化的追问还处在比较彷徨的时期，但是这种探索是我国设计师力求实现自我突破的重要一步。然而，这种转型在将来还会进一步发展，文化和设计风格将逐渐发展成为适合我国当下的一种量身制作的体系。这套体系将会成为我们国家独特的设计理念的源泉。

目前我们在探索这样一条明晰的道路。各种优秀的设计思想都将成为我们通往彼岸道路的坚固砖石。时至今日，中国的室内设计是该到了走出混沌的时候了。2012 广州国际设计周“装饰艺术趋势发布会”上以邱德光、梁志天、吴滨为代表的一批中国顶尖设计师都在这一关键时刻亮出了自己的风格力作。邱大师从《易经》当中寻找灵感，将明式家具时尚化，创作出梦云、风云等 T.K. 系列家具。梁大师联手 FENDI 、VERSACE 、ARMANI 等高端时尚品牌从事家居设计，将各品牌的精髓与风格完美融入时尚家居之中。当豪宅不再是炫富，当奢华融入生活，家便是一种质感的存在。通过材料的选择（例如使用纯手工艺高端定制的工艺品、绿色节能的新型材料装饰品）、颜色的搭配，线条和光线的运用，以及不同风格的家具和艺术品的融合，关注细节、品质，对时尚健康的生活态度的追求，从视觉、听觉、嗅觉、触觉和味觉上让业主感受到尽兴，体验到一个完美并充满情感的家，这些都体现了对豪宅的再定义。每一步不仅是前一阶段的结束，更是后一阶段的开始，它将为我们新的设计风格拉开序幕。设计师在这一过程中将人生感悟融入其中，富有创意又不拘一格的设计赋予了居住空间蓬勃的生气和温馨的情感。他们这一将生活的体验融于设计、创造人性化的奢华空间的过程，也让他们在享受生活的同时享受设计。

梁志天大师经常说的一句话就是“享受生活，享受设计”，对此他表示“设计源于生活，只有懂生活的人才能抓住设计的精髓，设计又能引领生活，真正优秀的设计会为人们享受生活创造更多的可能性。”我们不是自然的主宰，但我们却可以成为我们自己生活的主宰。愿更多好的设计，带给我们更多美好的生活。

XURIOUS MANSION

CONTENTS 目录

REDEFINITION OF I

UXURIOUS MANSION

YUANHANG VIOLET LAKE INTERNATIONAL GOLF VILLA

远航紫兰湖国际高尔夫别墅

- Designer: TSUN FONG (Hong Kong Fong Wong Architects & Associates)
- Location: Chengdu
- Area: 720 m²
- Photographer: Jiang Guozeng

- 设计师：方峻（香港方黄建筑师事务所）
- 项目地点：成都
- 项目面积：720 m²
- 摄影：江国增

Italian neoclassical style dominates the project with rich 18th century details and expressions, whereas resident-friendly layout provides enchanting visual effects and better spatial interactions, allowing unobstructed movements that contribute to enjoyment. The majestic living room, embellished with exquisite ornaments and stripes of black and white, is a mixture of the classic and modern as well as a delightful visual experience.

我们试图诠释出意大利新古典的格调，整个设计以更加温馨的手法营造出引人入胜的视觉效果，使人与居住空间有更多的互动。新古典风格有许多意大利十八世纪的装饰元素来衬托出空间的精神，也有更多的表现形式。起居空间的设计给人以尊贵感，黑白相间的条纹呈现出的是一份既现代而又古典的气韵，中间的装饰与点缀使空间呈现出更加良好的视觉效果。而空间的动线设计让人们身处空间中更加随意而自然，不至于受到过多的干扰，个中体验不言而喻。

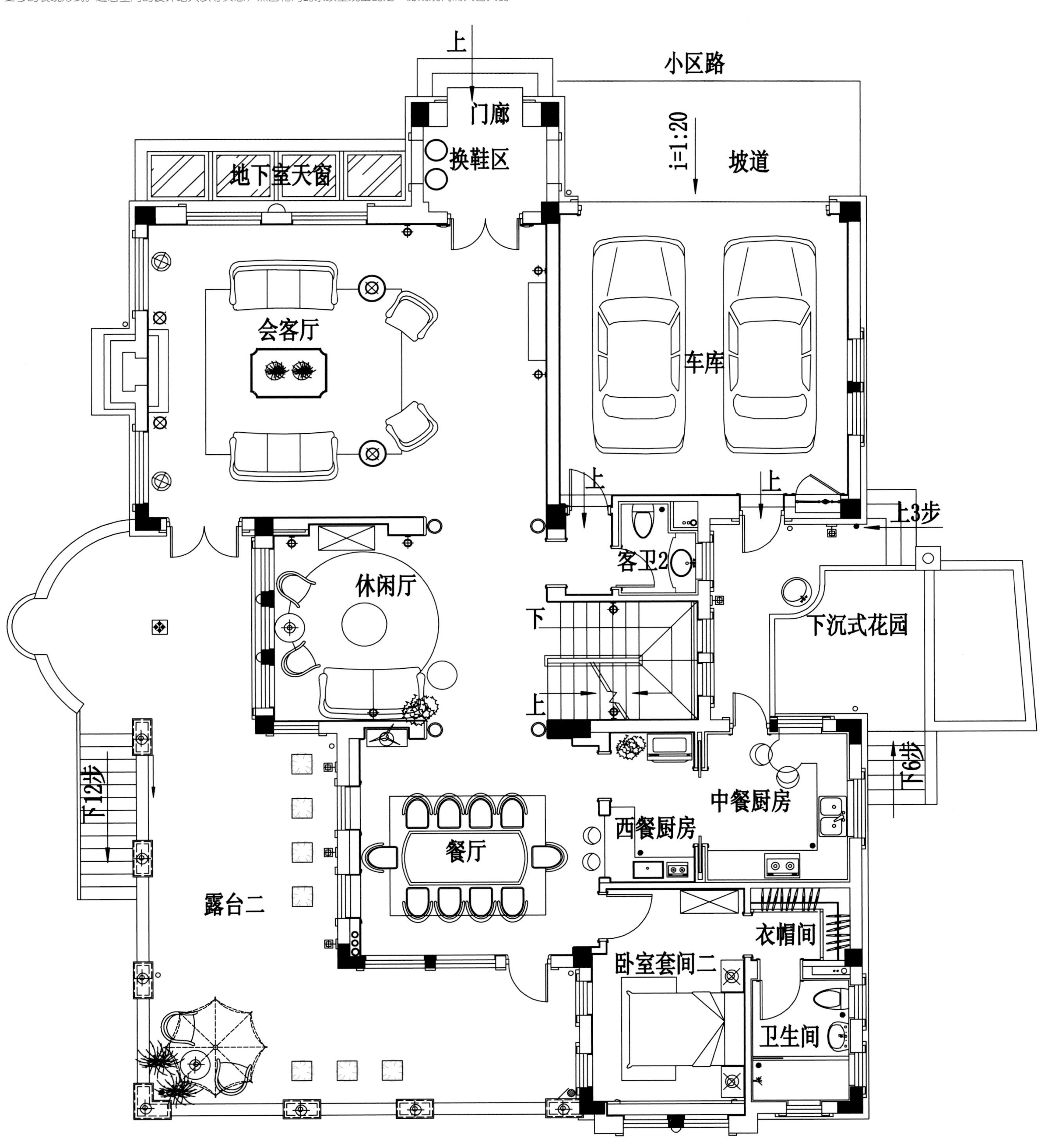

Plan 平面图

Wok

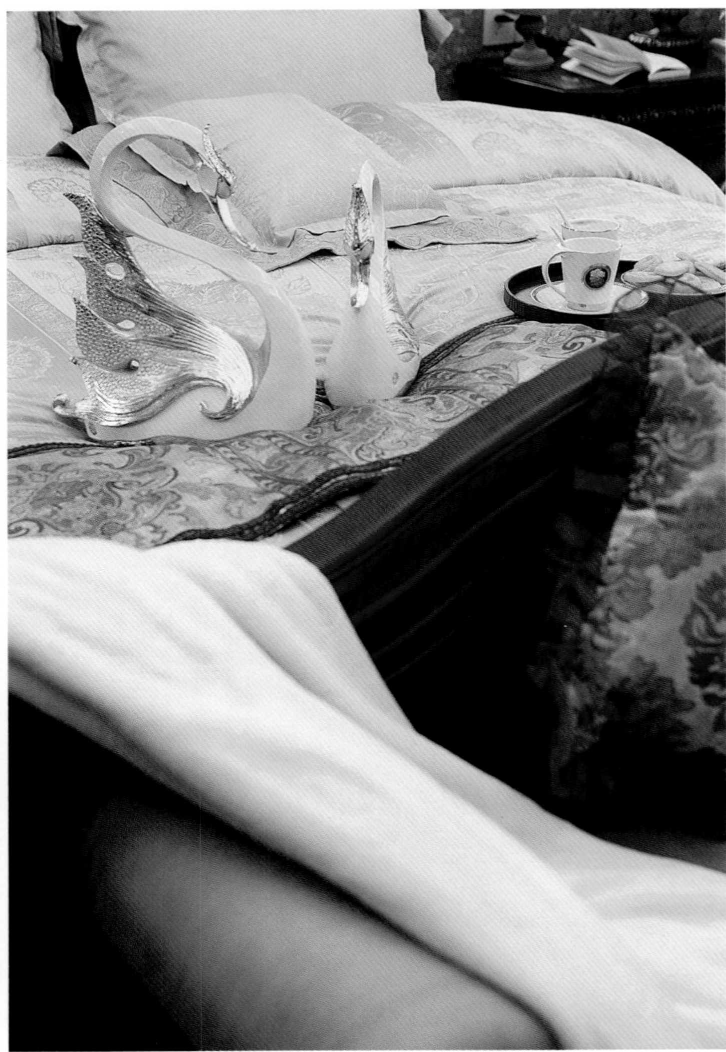

Arletta

ORIENTAL GRACE LUXURIOUS VILLA

东方美郡奢华别墅

- Designer: Wu Xuqiang
- Location: Guohua Oriental Grace Living Quarters, Jinan
- Area: 400 m²

- 设计师：吴绪强
- 项目地点：济南国华东方美郡
- 项目面积：400 m²

The deluxe villa, situated in Guohua Oriental Grace Residential Quarters, is the exact incarnation of splendor. The designer discontent with mindless duplication of the past and influenced by multivariant ideas and provides the residence with graceful outlines and grains. The project, therefore, is a modern breath mingled with the classic. Luxury and vogue come from a contradiction, where chaste modern furniture is subtly deployed in a classic atmosphere. The final effect of elegance and low-profile luxury is attributed to the designer's profound perceptions, mastery of materials and careful examination of limits. Every detail, the floor, walls, the ceiling, chandeliers, furniture, and miniature ornaments, etc, is attended to with consonant hues, proportionate shapes and fine materials, while every element is well coordinated during conception, contributing to the residence's enchanting magnificence.

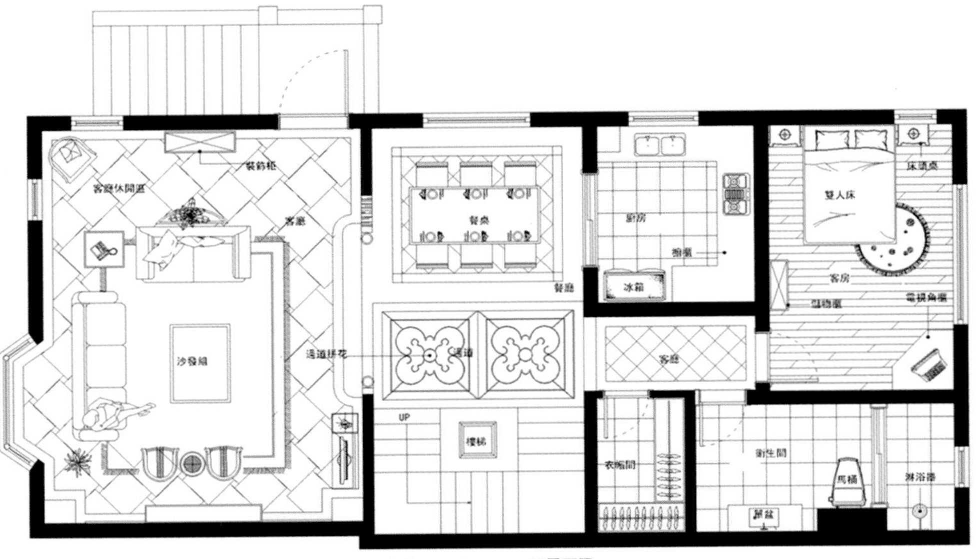

A floor plan　一层平面图

SPANISH BAR
SPANISH
ANISH
BAR
ANISH
BAR

The two floor plan　二层平面图

本案位于国华东方美郡的别墅区，这座豪宅府邸深刻地诠释了欧式新古典风格的雍容华贵。在多元的影响下，设计师并非简单地复制古典，而是将古典融入到现代中，创造出优雅的造型与纹路，体现了现代的材质与工艺之美。古典的装修氛围，搭配现代的典雅家具，用碰撞和矛盾的手法，融合出奢华的时尚感。设计师对材质的深刻把握，对古典生活的独特见解，对尺度的掌握和考量，无疑使设计语言充满时尚气息，使优雅与从容的奢华气质毫不张扬地一一呈现。设计师注重造型的搭配和颜色的协调，墙面、地面、顶棚以及家具陈设，乃至灯具器皿都精心选材。一切元素在设计师巧妙的编排中各就其位，融入这古典奢华的场景之中，形成一派高贵雍容的气势。

JIANBANG VALLEY OF FRAGRANCE I

建邦原香溪谷 I

- Designer: Yue Meng (Imaging Space Planning, Jinan)
- Location: Changqing District, Jinan, Shandong
- Area: 240 m²

- 设计师：岳蒙（济南成象设计有限公司）
- 项目地点：山东济南市 长清区
- 项目面积：240 m²

The Luxurious sample house situates itself in Jianbang Valley of Fragrance, where unadulterated Italian Tuscany style reigns. The finest of Tuscany is annotated by the structures of these Sumptous duplexes, villas and courtyards.

Thorough consultation contributes to the design motif. The resulting layout boldly embodies deluxe and passionate Tuscany features despite of limited space, while original fenestration and spatial arrangements are considerately retained for optimal effects.

Pleasant sunshine and exuberant greenery provide a hospitable atmosphere. A Chinese kitchen and dining room, a western kitchen, a bar and a miniature cellar divide the first floor, while grand arches integrate the otherwise separate functional areas, visually enlarging the commodious residence.

The second floor is divided into bedrooms and a studyroom while allowing for a wardrobe and a dresser, with additional space for unobstructed movements. The square bathroom especially challenges spatial arrangement, ingeniously solved by coordinating a fine basin, a massage bathtub, a shower and a second dresser, sufficient for routine chores.

1. 沙发
2. 半圆端景柜
3. 茶几
4. 电视机
5. 餐桌椅
6. 吧台
7. 吧凳
8. 床
9. 洗衣机
10. 矮柜
11. 洗手盆
12. 座便器
13. 浴缸
14. 电视柜
15. 玄关柜
16. 电冰箱
17. 酒柜

A floor Plan　一层平面图

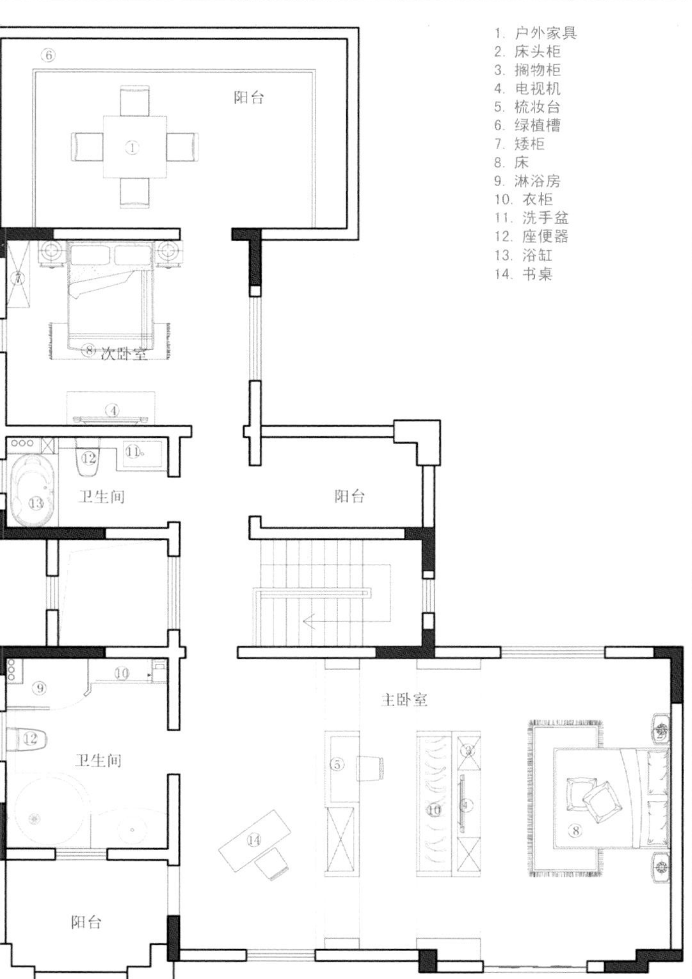

The two floor Plan　二层平面图

本案位于建邦原香溪谷纯正意大利托斯卡纳风格精品别墅区。复式结构的房子，精致奢华的别墅，方正的庭院加上完美的结构。这就是对托斯卡纳生活最好的诠释。

经过多方沟通后，设计师才大胆地将这套别墅定位成托斯卡纳中的奢华热烈风格，在并不是太宽绰的空间中，我们充分地尊重原有建筑在结构中的空间关系和窗与窗之间的，窗与景观庭院之间的对景关系。

首先让激情、健康的阳光和生机美丽的植物来为空间打底，然后设计师对一层大空间进行细化，分隔出中餐区、酒吧、中厨和西厨，以及小小的酒窖区，同时强调出大大的拱形造型，来弱化空间的区域感，真正做到了功能性强与空间宽敞的双赢。

二层是主人的起居区域，设计师同样分隔出了书房、衣帽间和卧室，人的生活动线被一一细化，同时方正的卫生间是重点要处理的区域，富贵的手盆、按摩浴缸、淋浴和化妆台，最基本的生活诉求在这里得到最完美的诠释。

JIANBANG VALLEY OF FRAGRANCE II

建邦原香溪谷 II

- Designer: Yue Meng (Imaging Space Planning, Jinan)
- Location: Changqing District, Jinan, Shandong,
- Area: 400 m²

- 设计师：岳蒙（济南成象设计有限公司）
- 项目地点：山东济南市 长清区
- 项目面积：400 m²

Upon entering, elegantly outlined arches rich in Italian features are immediately encountered, whereas perfect unification of chaste and dense hues provides the residence with a unique touch. An open western kitchen is completed with a round dining table, as a nod to Chinese culture. The impressive arches also offer views to the reception room furnished in traditional Italian style, where a grand fireplace and fine ornaments contribute to luxury and grace. Guest will leave wondering at the householder's sophisticated tastes.

The adjacent living room and home theatre enjoy favorable privacy, where a bar awaits the householder and friends with robust wines a Chinese chessboard and an appealing way to savor life.

A novel pitched roof adorns the master bedroom, demonstrating the inartificiality, magnificence and warmth of Tuscany, where slopes covered with exuberant groves and lakes reflecting subtle architectures are entwined with lovely sunshine and wind in the leaves. Join the designer in Tuscany, a wonderland amidst the bustling cities, where love lingers and dreams unfold.

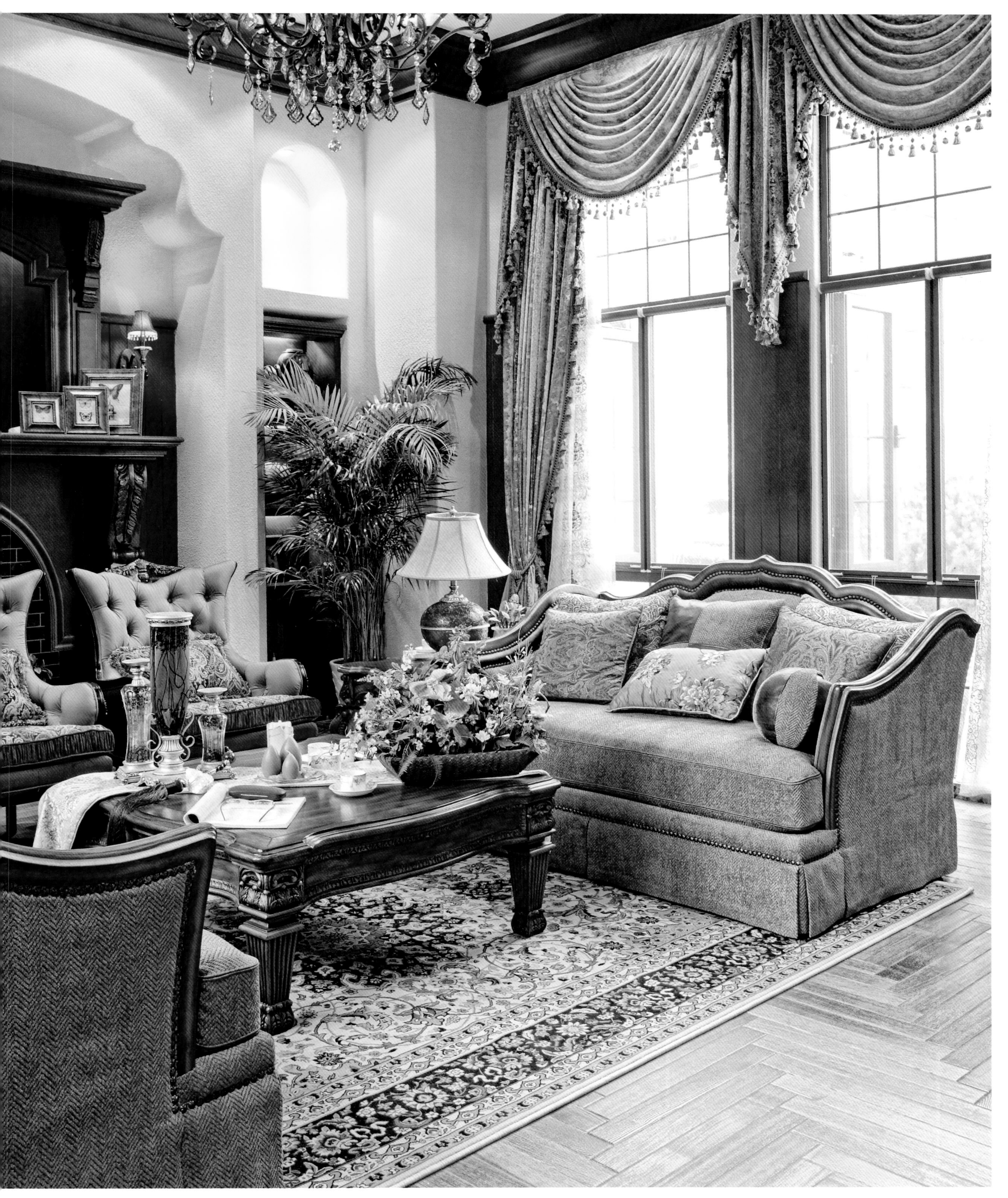

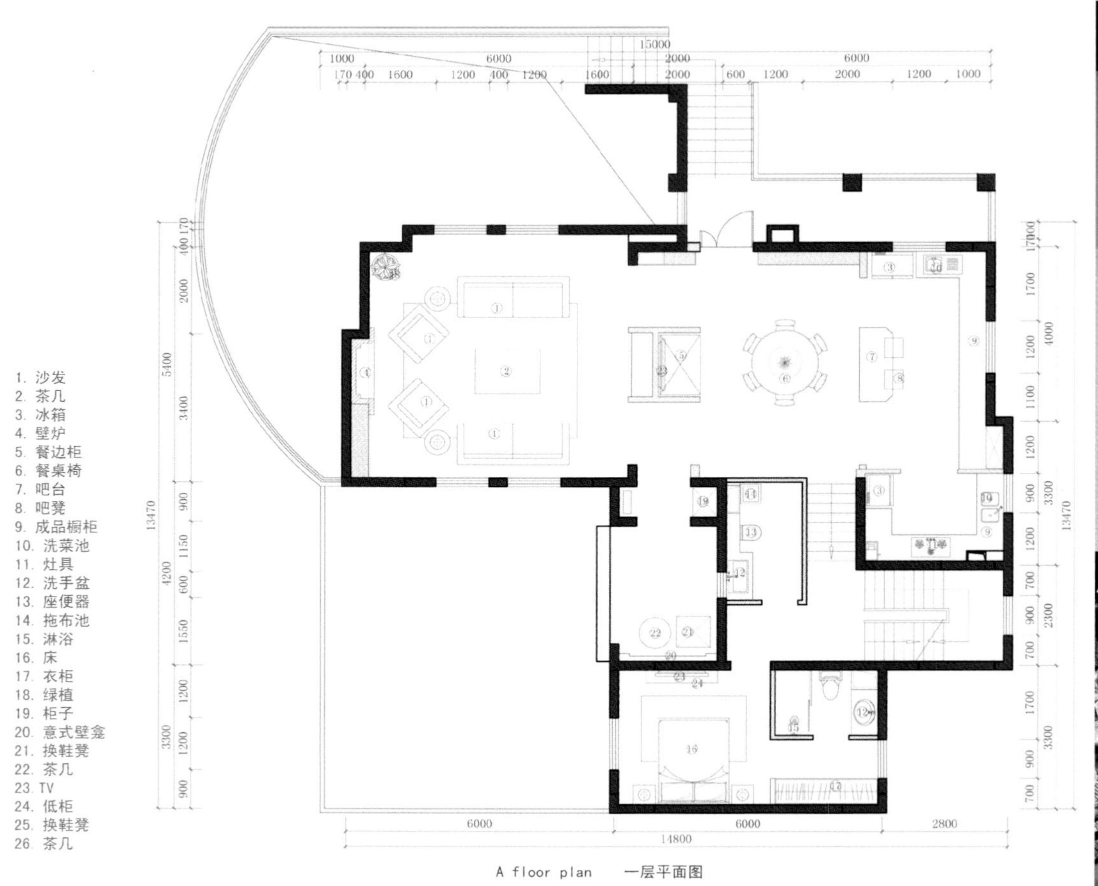

A floor plan　一层平面图

从入户门走进室内，映入眼帘的是标志性的拱门，优美的造型线，散发出来自意大利的建筑之美，淡雅的墙壁与浓郁而厚重的色彩结合得恰到好处。餐厅与西厨遥相呼应，圆形餐桌在此更加符合中国人的传统习惯。穿过高大厚重的拱形门，是主人接待客人的厅堂，高大的壁炉和传统的意式家具以及精美的饰品摆设，质朴中彰显出大气与华贵，体现出主人极高的生活品位。

紧挨着的是相对私密的起居室与影音室，这里设有吧台，主人可以携老友在此品尝浓郁甘醇的红酒，或者一起来几局象棋，感悟一下棋局人生。

主人卧室，别致的坡屋顶迎合了托斯卡纳传统质朴的形式，但又不失华贵与温馨。这就是托斯卡纳，这就是城市中的田园，那青山，那绿树，那湖泊，那质朴的建筑，还有那暖暖的阳光，微微的清风，来到这里，爱上这里，留恋这里……此刻，我在托斯卡纳！你呢?

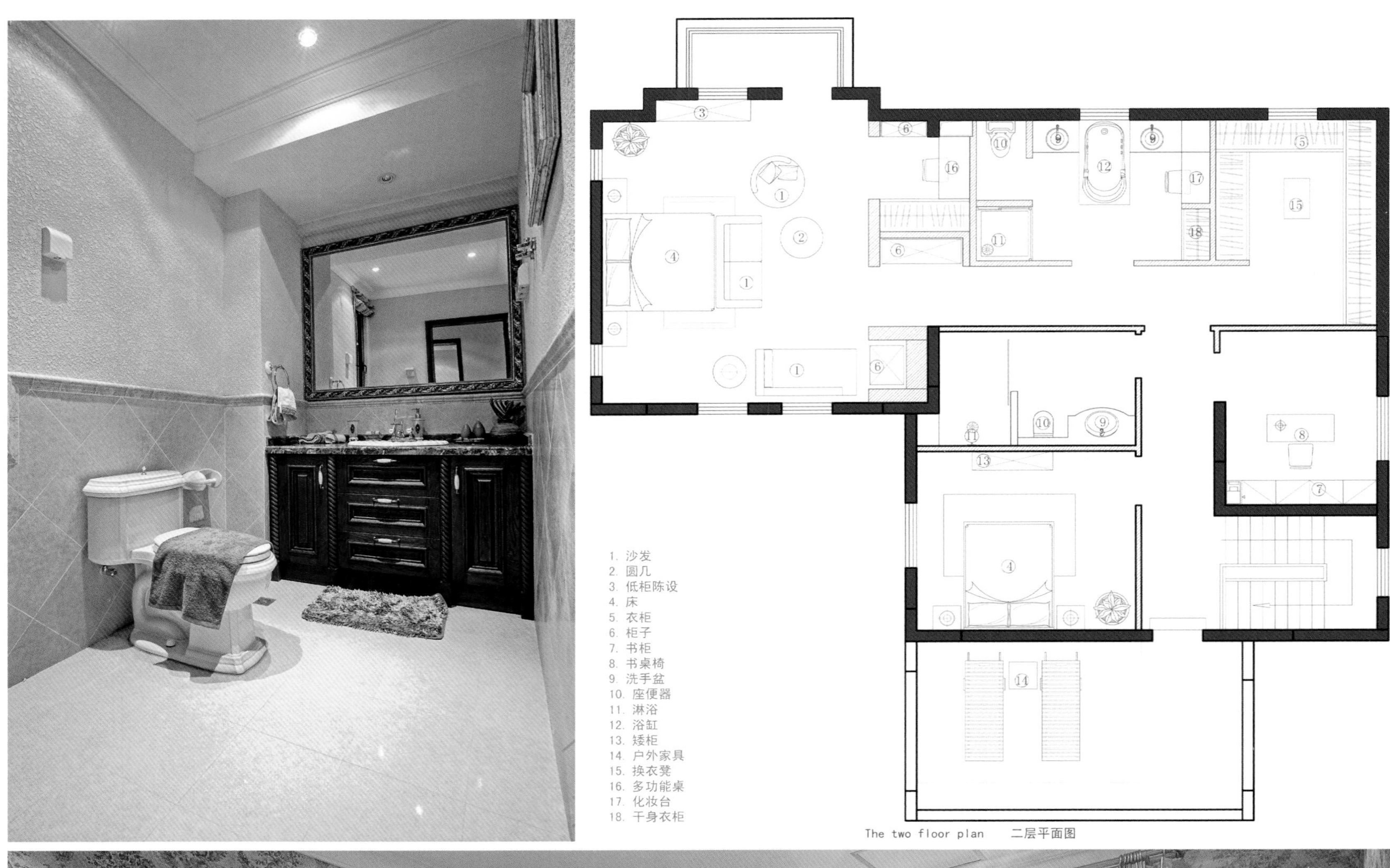

The two floor plan 二层平面图

ANHUI HUA DI
ZI YUAN

安徽华地紫园

- Designer: Wang Guan (Matrix Interior Design)
- Location: Anhui
- Area: 375 m^2

- 设计师：王冠（矩阵纵横室内设计）
- 项目地点：安徽
- 项目面积：375 m^2

Wine proves a romantic design motif for an innovative sample house. Combining a western house's first floor and basement, the resulting duplex is completed with a garden level. Sishuiguitang (a traditional residence native to regions south of Yangtze River) inspires the design concept, as neighborhood water is allowed inside, contributing to a waterfall running down from the upper to the lower floor, clearly visible upon entering. The residence'resemblance to a grand mansion is attributed to subtle partitions and a perception of depth, promising a brandnew view with every step. The western house's obscure lower parts, therefore, is ingeniously transformed to rival a villa.

基于洋房的一楼以及带有下沉庭院的复式结构，设计师以红酒为主题演绎全新的亲地别墅样板房。入户即开始通过泗水归堂的概念将整个小区园林的水系引入本户型，通过流水瀑布的手法贯穿两层空间，并在空间关系上强调了登堂入室的豪宅观念，同时层层借景，步步引人入胜。在一个原本只是洋房的底层空间打造出一个比拟别墅的高品质空间。

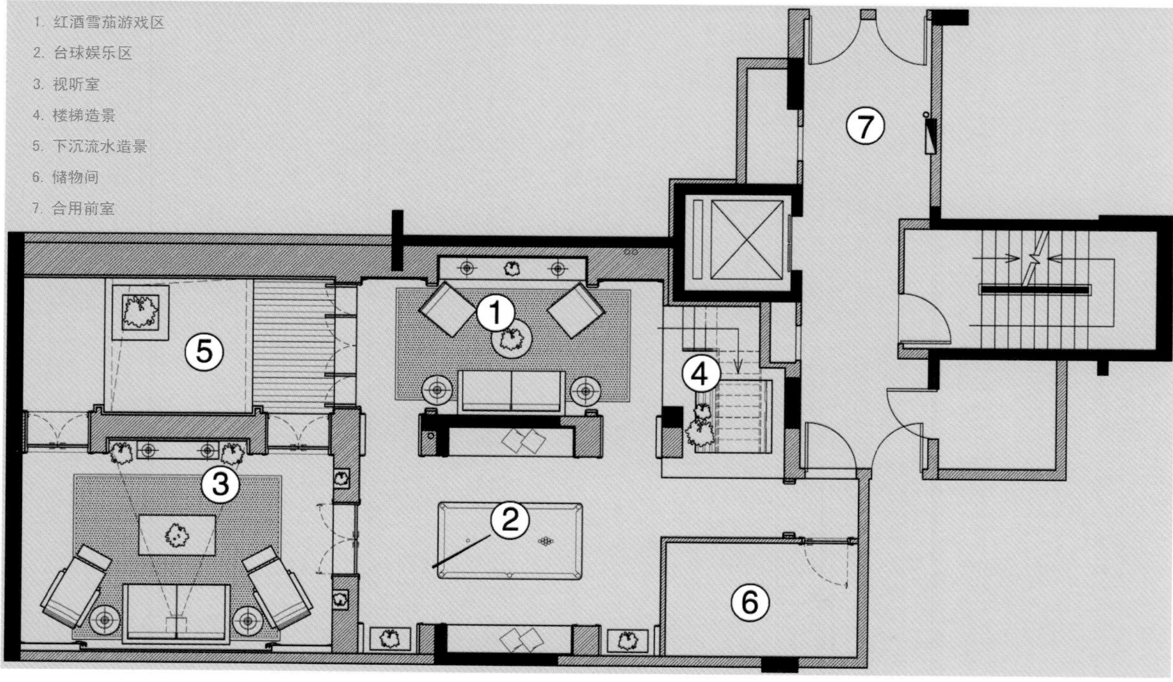

Basement plan　地下室平面图

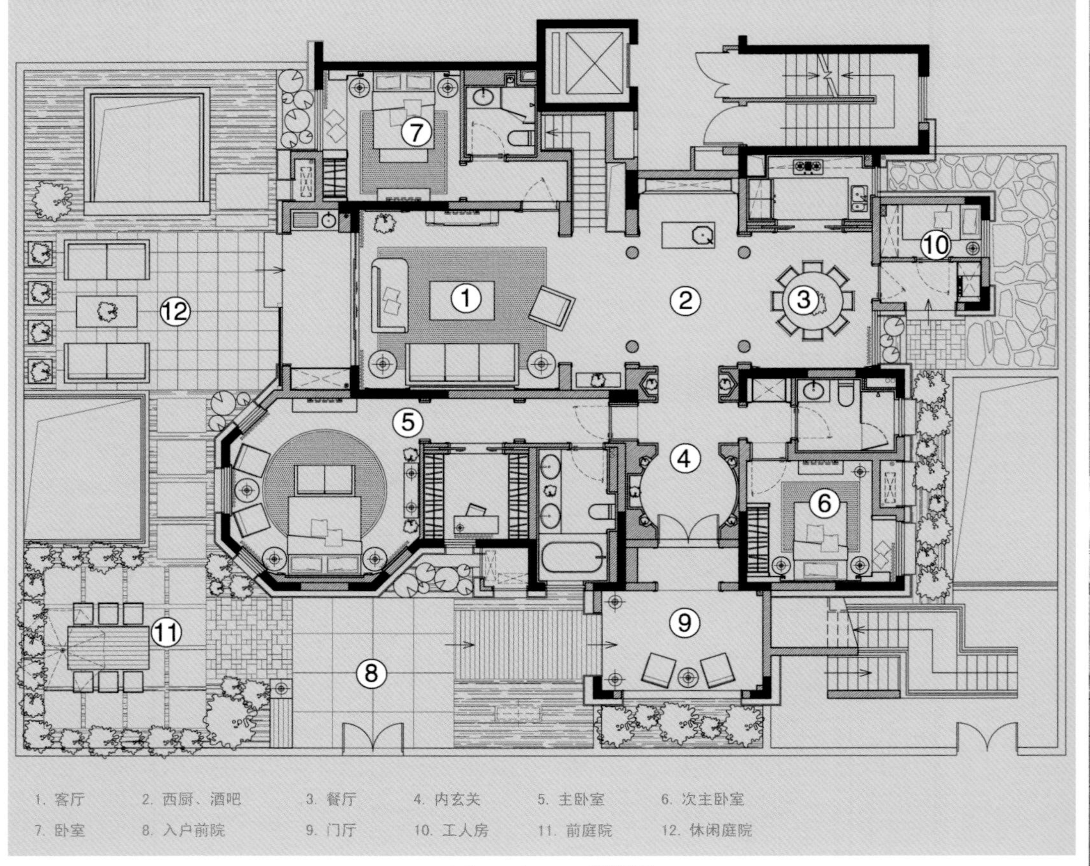

A floor plan　一层平面图

GEMDALE GRAND SCENERY

金地湖山大境

- Designer: Li Jianming (Scale Design Co., Ltd., Dongguan)
- Location: Huangjiang, Dongguan
- Area: 590 m²

- 设计师：李坚明（东莞市尺度室内设计有限公司）
- 项目地点：东莞黄江镇
- 项目面积：590 m²

The Art Deco residence is tailored to its overseas Chinese householders, who are equestrianism fans. The project reserves exquisite ornaments and considerate lifestyle details, expressing refined aesthetic appreciation of the modern aristocracy. Distinct functional areas are reasonably allocated to four floors, which the underground floor is used for reception and recreation and the rest of the building for daily life. Displays and decorations carefully attend to the householder's preferences and breathe life into the residence. High-end building materials are synonymous with luxury, while neat geometric patterns and shades integrate the entire layout, where grandeur pervades.

凯仕乐
KASTTOW

花园
楼梯间
上
下
厨房
采光井
上二级
会客厅
餐厅
过厅
公卫
长辈房
下
Plan　平面图

THRONE ROOM

本案采用 Art Deco 风格的装饰元素，并赋予喜爱马术的华侨家族作为故事背景，从精致的装饰到生活的细节均体现出屋主对高雅品位的理解，引领奢华贵族的生活方式。项目共有四层，地上三层为起居空间，地下一层为会客、娱乐空间，设计师清楚分配给每一层空间不同的机能，并结合屋主的喜好与品位用陈设布景的手法呈现有灵魂的居家空间。运用高级建材构筑空间奢华品质，以规则的几何图形组合空间和色彩将整个调性统一起来，整个空间散发大气豪迈的气质。

POLY GOLF VILLAS

保利高尔夫别墅

- Designer: Wu Changyi (Chongqing Maisuo Interior Design Co., Ltd.)
- Location: Chongqing
- Area: 650 m²

- 设计师：吴昌毅（重庆麦索装饰设计有限公司）
- 项目地点：重庆
- 项目面积：650 m²

Inspired by classical European literature and rich in dainty hues and materials, this residence is synonymous with luxury and majesty. Splendid ornamentation extends to every detail. Fine marble pillars, gorgeously patterned carpets and magnificent chandeliers restore classic European essence, leading to the overwhelming visual impact and innate exquisiteness.

Classic European glamour resides in every marquetry, as exemplified by the project's refined furniture. The pure white master bedroom refreshingly changes the atomsphere. The residence also reserves a complete collection of functional areas, providing a sophisticated European symphony pervaded by fashion and luxury.

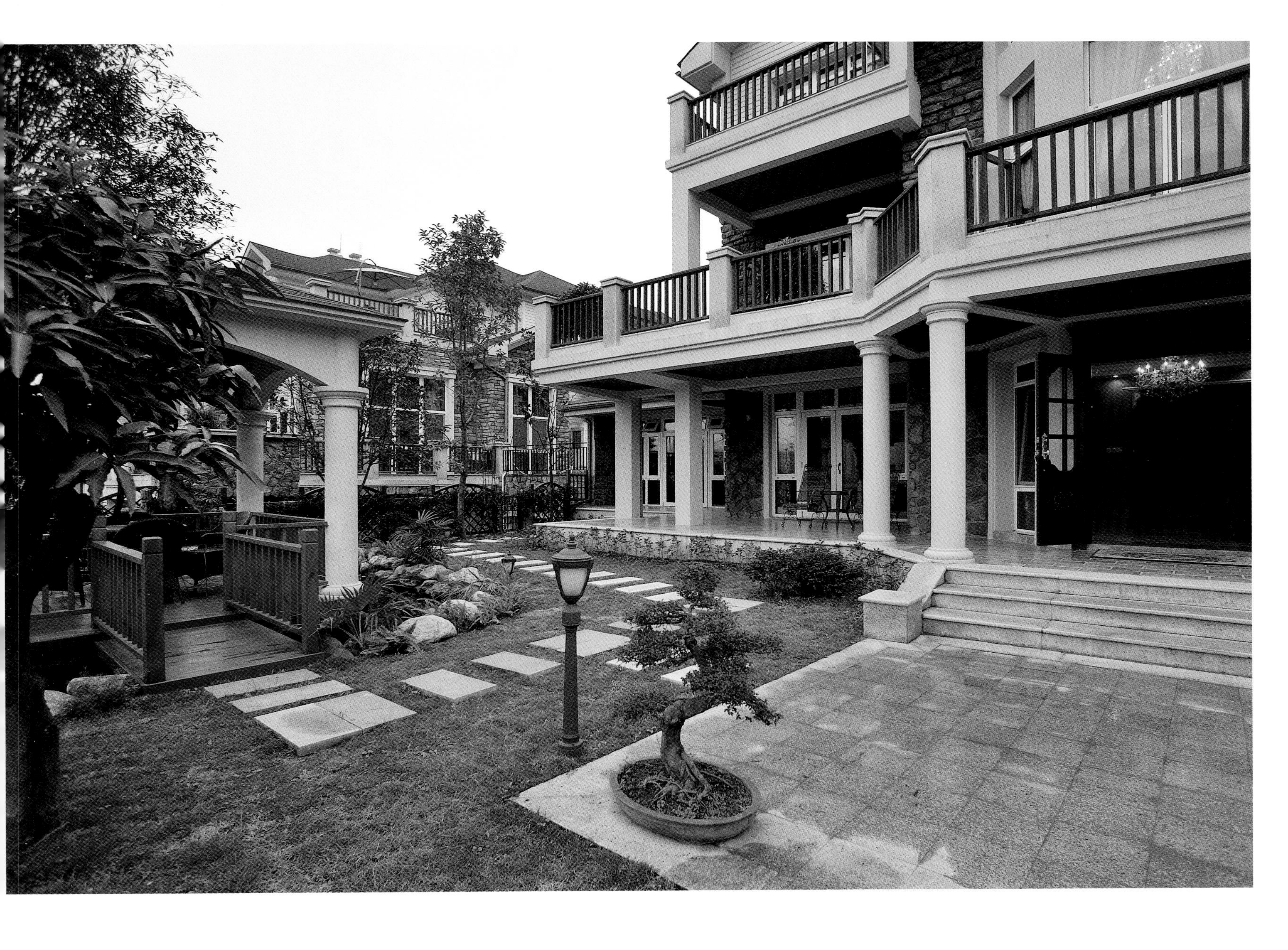

本案意图以古典欧式风格表现，透过时尚的色彩和材质，彰显着空间的雍容大度、豪华气派。进入空间，迎面的就是满室金碧辉煌的装饰，大理石打造的罗马柱、颜色鲜艳图案奔放的地毯。华丽的水晶吊灯，这些传统的强势元素，以及强烈的视觉冲击力，烘托出古典欧式的精髓。他们配衬着生活的每个场景，以典雅的方式默许着奢华的态度。

古典欧式的华丽感是以细节处的精雕细琢来成就的，这在各式家具的繁复雕花中可见一斑。卧室则以白色为主调，呈现了较为清新爽朗的空间氛围，使人产生耳目一新之感。注重装饰的同时，设计师还兼顾功能区的齐全，在这个奢华的空间中，演绎一首欧风弥漫、浓墨重彩的时尚交响曲。

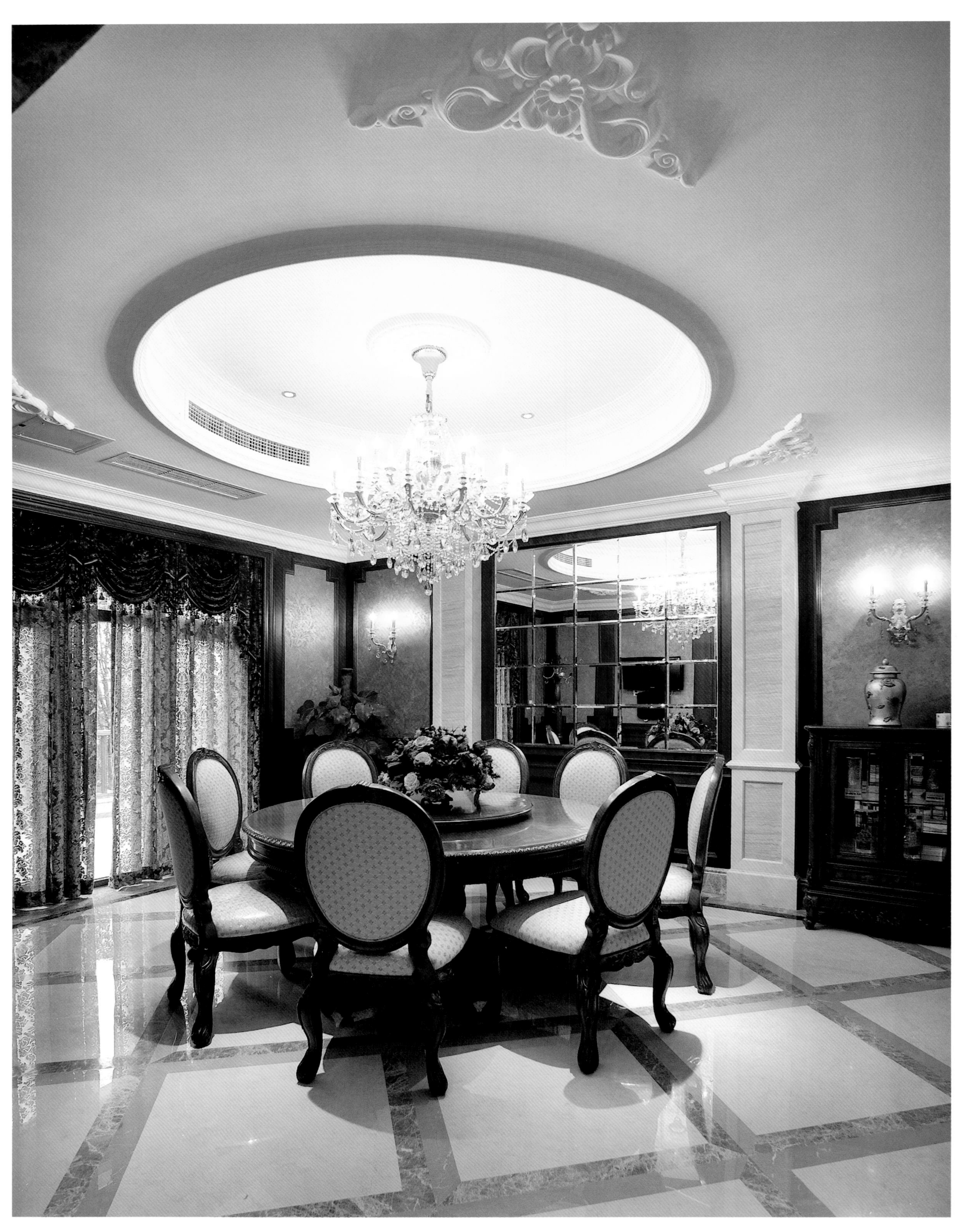

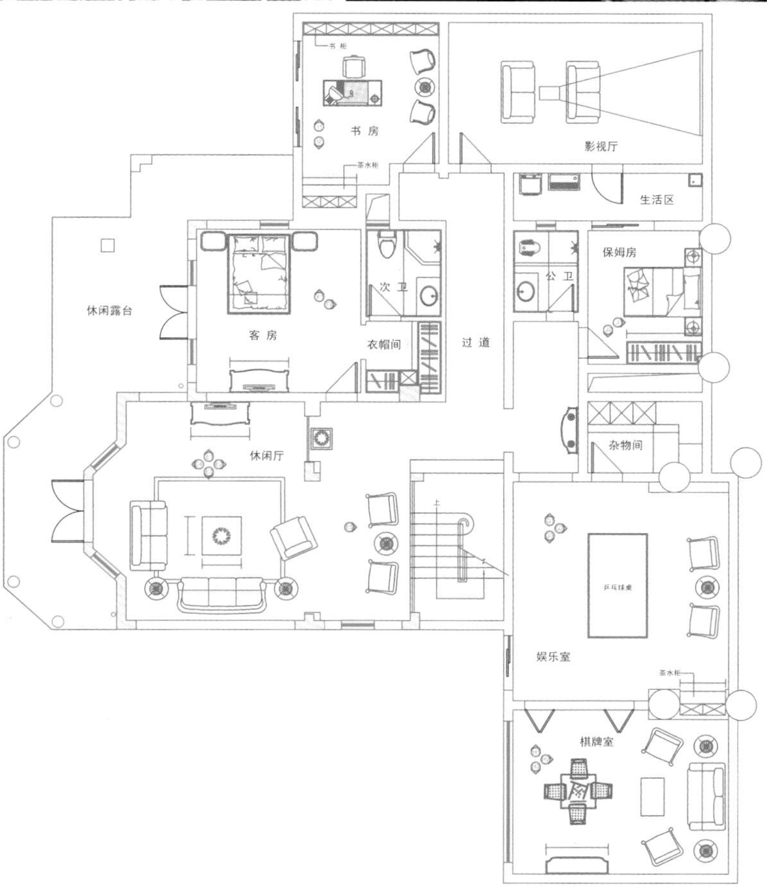

Basement plan　地下室平面图

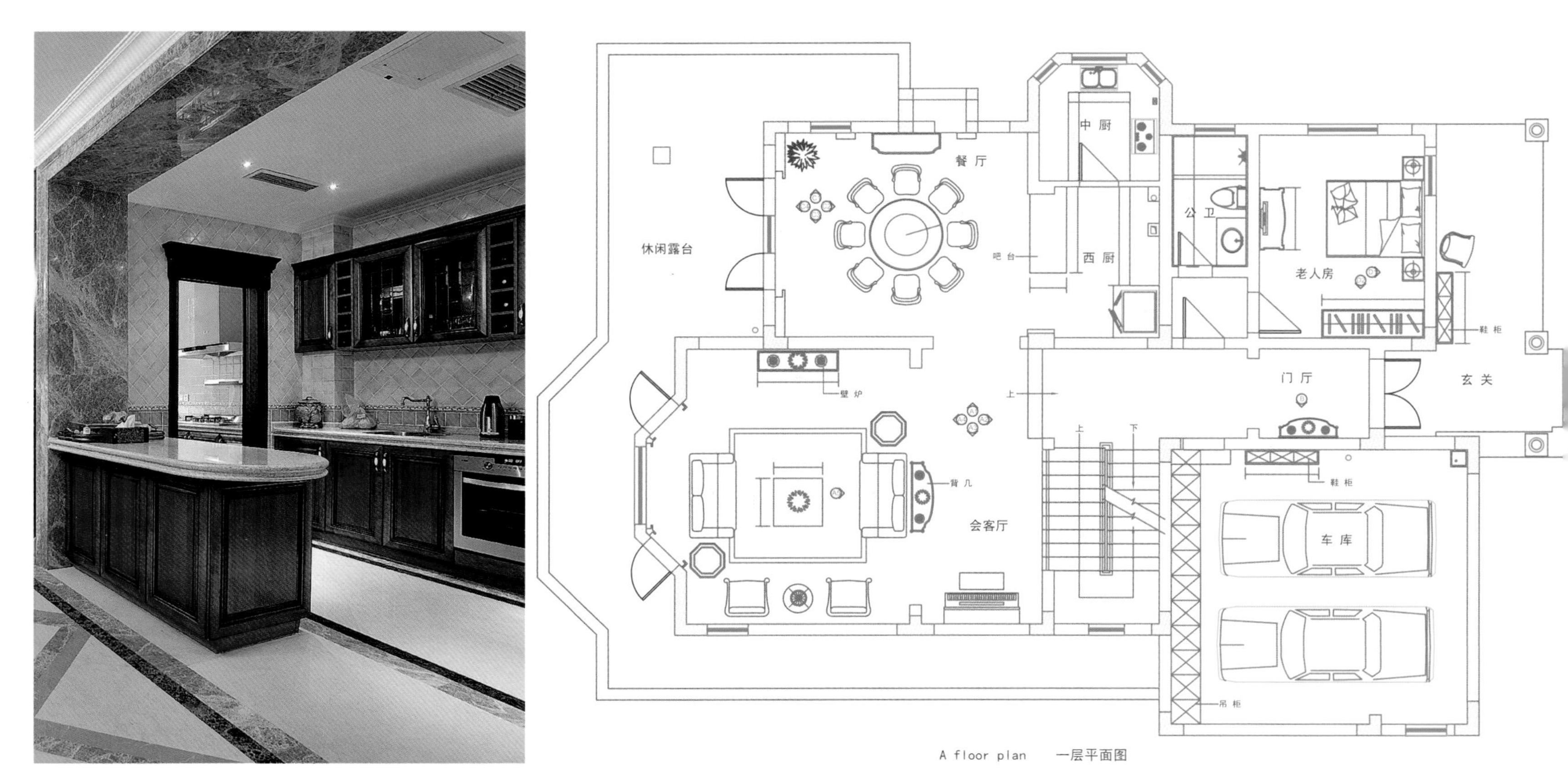

A floor plan　一层平面图

2010·2011 世乒赛专用台
DHS

GEMDALE'S LAKE CONSTANCE

金地博登湖亦居

- Designer: Li Jianming (Scale Design Co., Ltd., Dongguan)
- Location: Tangxia, Dongguan
- Area: 810 m²

- 设计师：李坚明（东莞市尺度室内设计有限公司）
- 项目地点：东莞市塘厦镇
- 项目面积：810 m²

Targeted questionnaires reveal the locals' admiration for alleged European architecture, whose overly elaborated features, however, unpleasantly challenge the reserved Chinese aesthetics. Attending to locals' taste, the designer opts for simplified European style. However, unreconciled to the norm, he embellishes wall frames and the drawing room ceiling with deluxe silver foil, an ingenious compromise between traditional and simplified European genres. Artificial leather and golden foil are innovative touches of luxury, broadening tradition range of European materials, e.g. marble, fabrics, paint and wallpapers. etc.

根据对当地居住人群的调查，相当多的一部分人比较向往欧洲居住的文化生活。对于欧洲风格来说，简欧风格更为清新，也符合中国人内敛的审美观念，所以简欧风格是为这类消费群体量身订做。本案颠覆传统风格上的简欧，在客厅以及墙身的线框和客厅天花上加以奢华的材质——银箔，使得作品拥有一种比传统简欧更奢华的气质。传统的欧式用材常常使用大理石、布衣、油漆、墙纸等等，本案新添加了比较奢华的材质，如人造皮革、金箔等等。

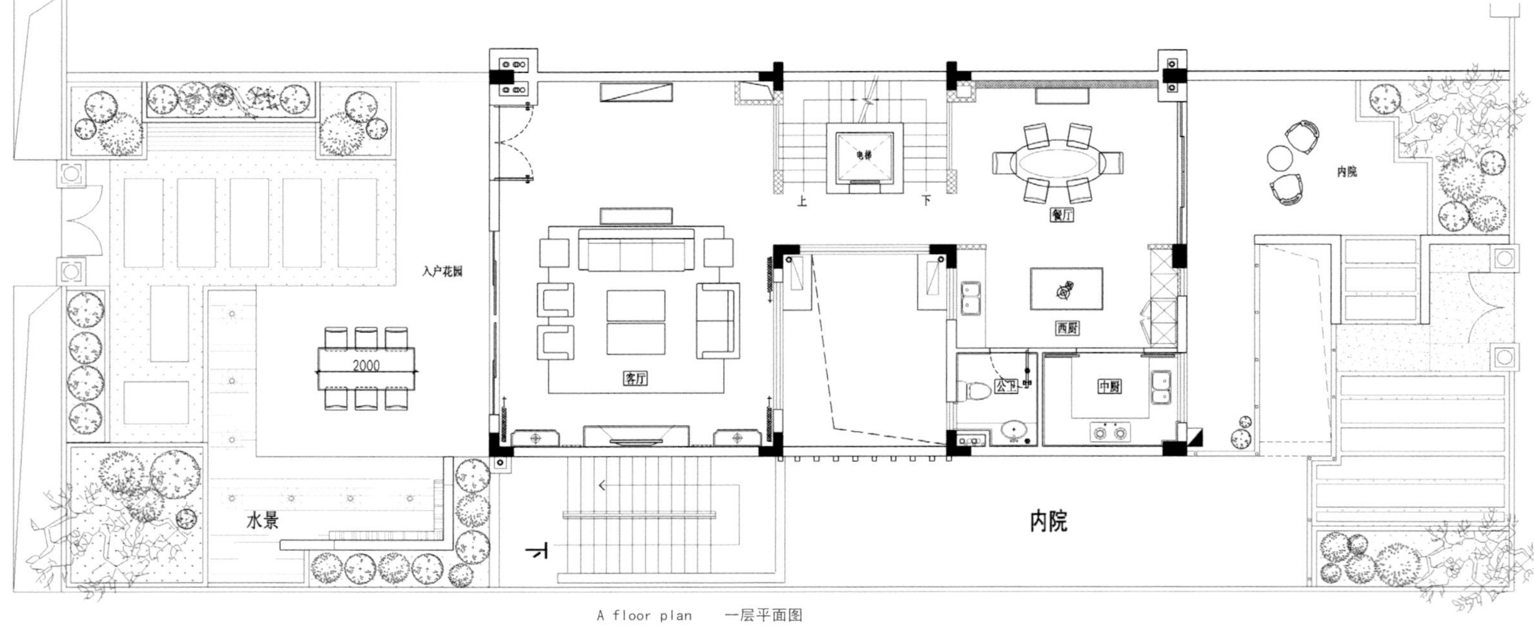

A floor plan　一层平面图

FASHION OF THE 70s

BEST-LOVED

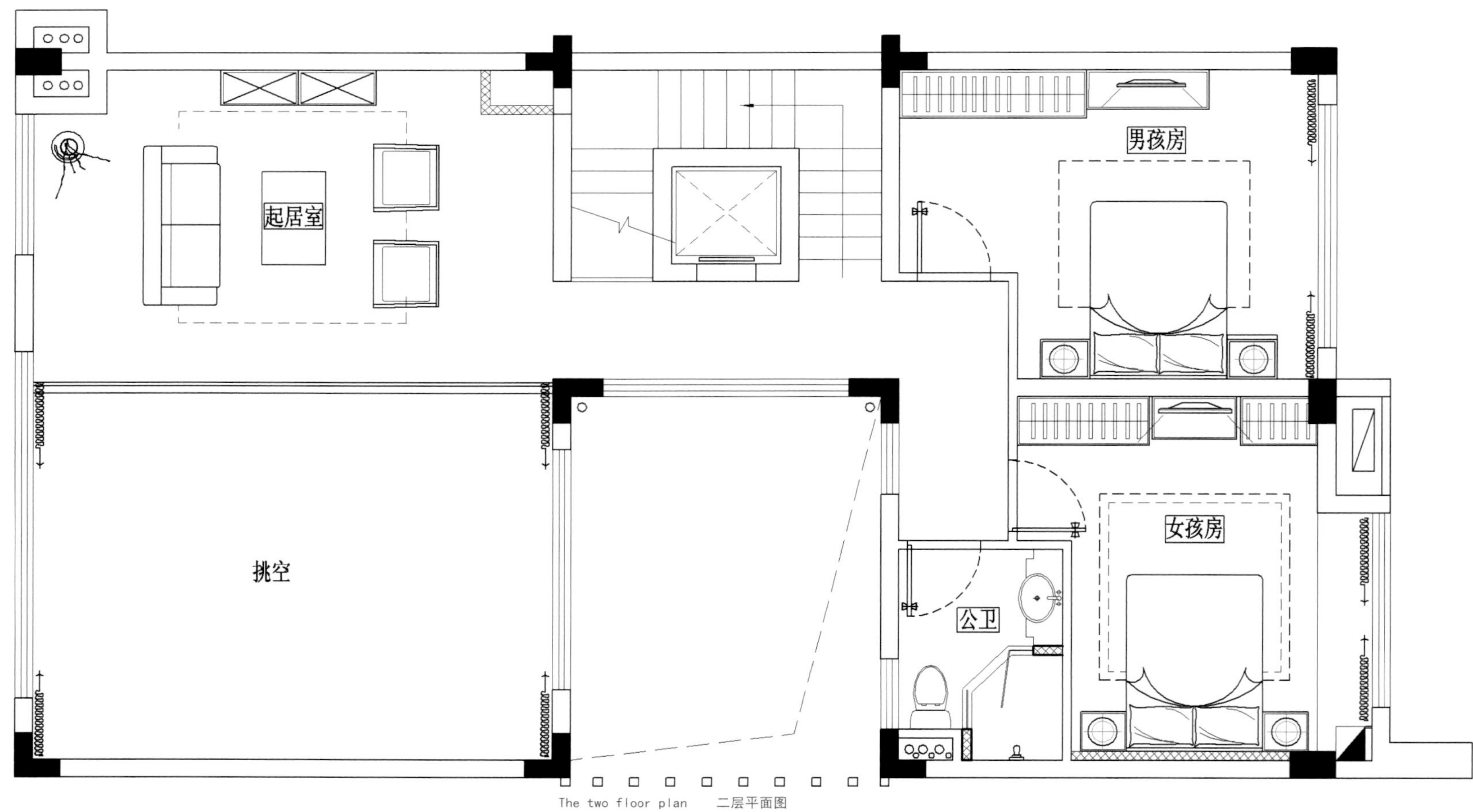

The two floor plan　二层平面图

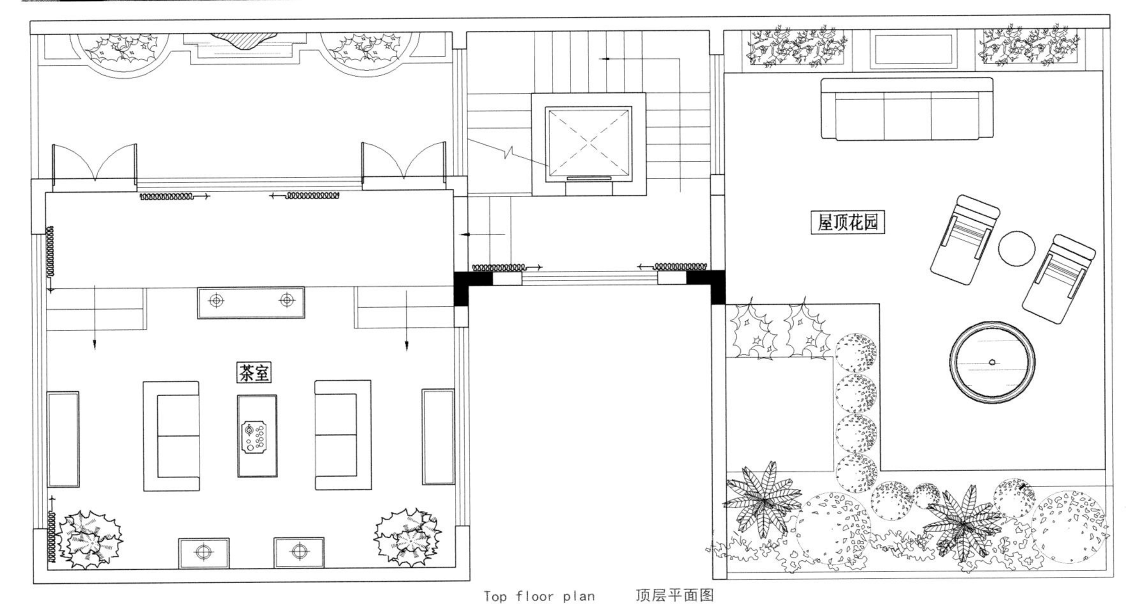

Top floor plan　　顶层平面图

TIAN YUE BAY

天悦湾

- Designer: Wang Guan
- Location: Shenzhen
- Area: 600 m²

- 设计师：王冠
- 项目地点：深圳
- 项目面积：600 m²

In this case, beige is matched with gold within the framework of French classic lines. From the whole to parts, the design is more like a diversified way of thinking, which combines the classical romantic feelings with modern people's needs towards life as well as integrates sumptuousness and elegance with fashion and modernity.

On the facade, the classical pure white line wallboard is matched with the wallpaper with modern patterns. The designers use the stone parquet on the ground and adopt the crude texture and natural colors of stone to represent the sense of quality of the entire space, which makes the luxury, high-grade and taste of the living room and the bedroom flow without reservation. In terms of furniture collocation, the solid wood furniture and panels as well as the lacquered surface of furniture with a semi-closed effect not only can fully show the texture of veneer, but also make people feel the smoothness and neatness behind the lacquered veneer when they touch it. For accessories decoration, the designers adopt white, light yellow, dark brown and other colors as the keynote and blend a little gold to make the color look bright and generous, so that the whole space give people an open, tolerant and extraordinary manner.

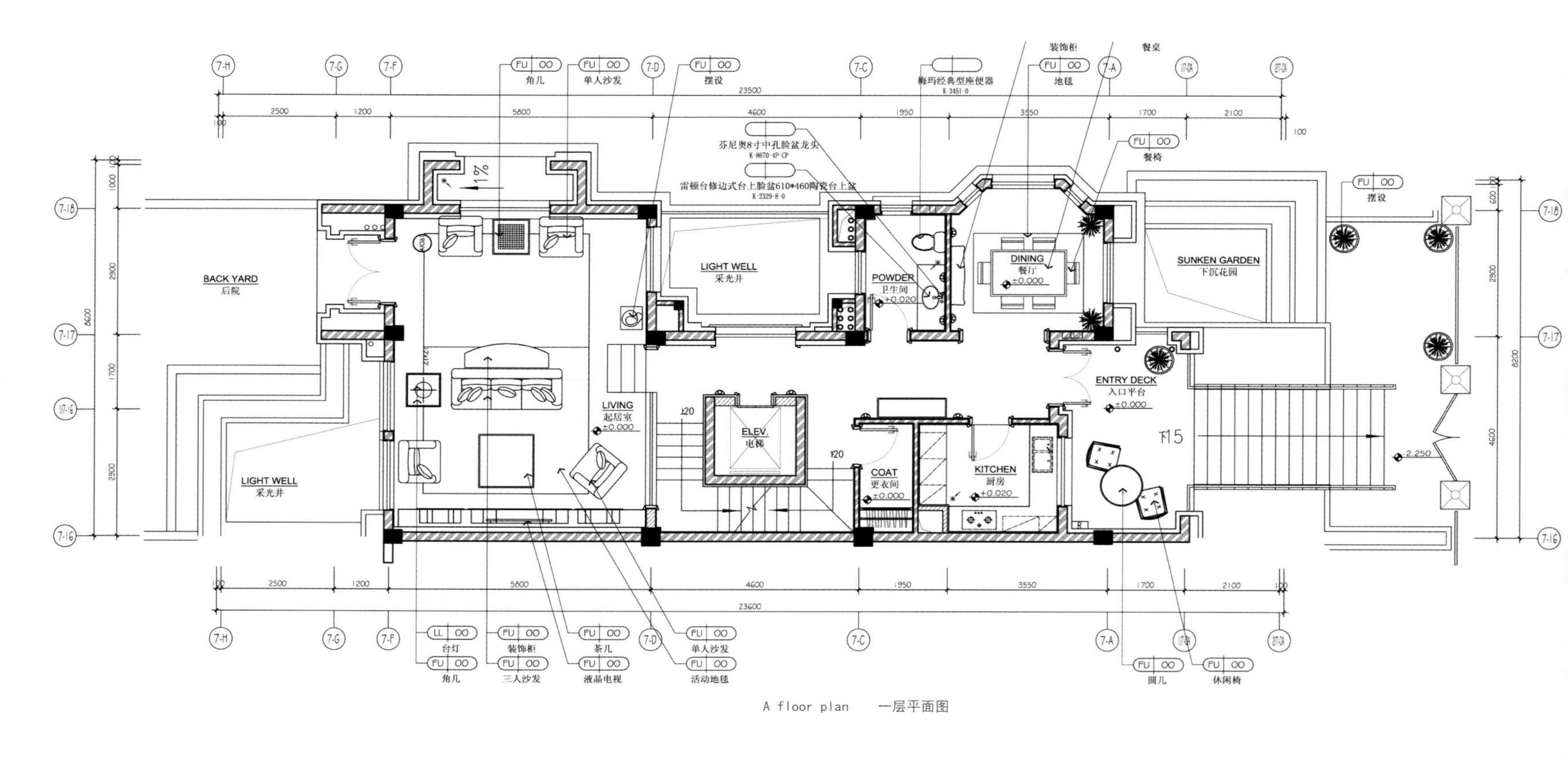

A floor plan 一层平面图

本案的设计在法式经典的线条骨架中，以米色搭配金色，从整体到局部，更像是一种多元化的思考方式，将古典的浪漫情怀与现代人对生活的需求相结合，兼容华贵典雅与时尚现代。

在立面上用古典纯白色的线条墙板，搭配着现代纹样的墙纸。地面采用石材拼花，用石材天然的纹理和自然的色彩来体现整个空间的品质感，使客厅和卧室的奢华品位毫无保留地展现出来。在家具配置上，板木结合的实木家具，家具漆面具有半封闭漆效果，不仅能将木皮的纹理尽情展示，在徒手触摸时，还能感受到油漆饰面后的光滑、平整。在软装配饰上使用白色、浅黄色、暗褐色等颜色为基调，少量金色糅合，使色彩看起来明亮、大方，使整个空间给人以开放、宽容的非凡气度。

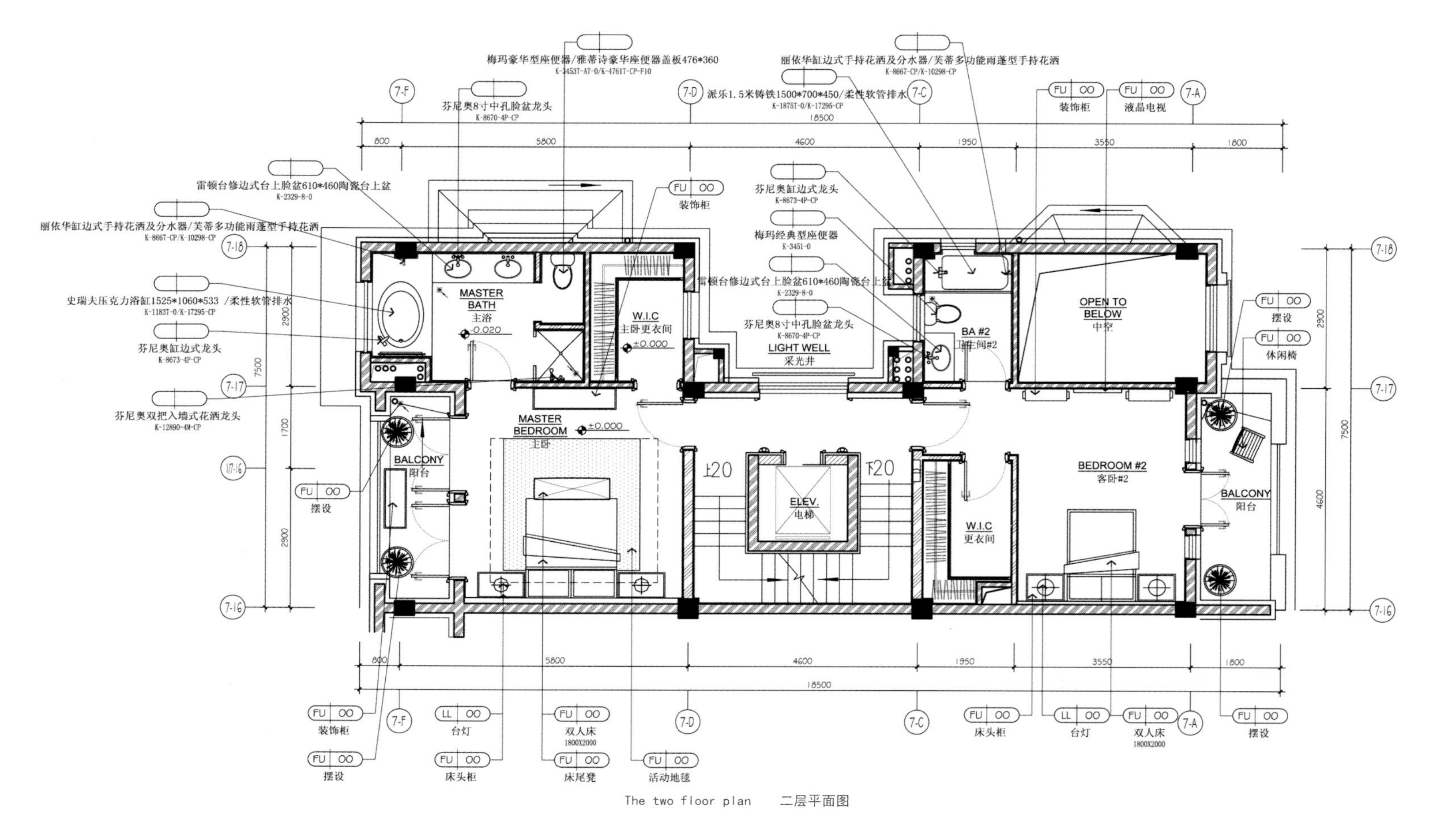

The two floor plan　二层平面图

ROMANCE-FINE

VILLAS

美墅馆

- Designer: Li Jianhua
- Location: Changshu, Jiangsu
- Area: 400 m²

- 设计师：李建华
- 项目地点：江苏常熟
- 项目面积：400 m²

The residence, rich in American style, corresponds with the householder's social status and unique lifestyle. It is a token of nostalgia, earnest love for life and resilient pursuit of freedom.

Integrating traditional American and local cultures, the project earns a style of its own, refining the residence's innate impressiveness. The entire interior is optimized and transformed to highlight a perception of depth, where a compound of dimensions satisfies the vision of the fastidious.

The design embodies nostalgia, romance, unrestraint and refined taste offering a place to forget the passage of time in the fragrance of tealeaves.

Arletta

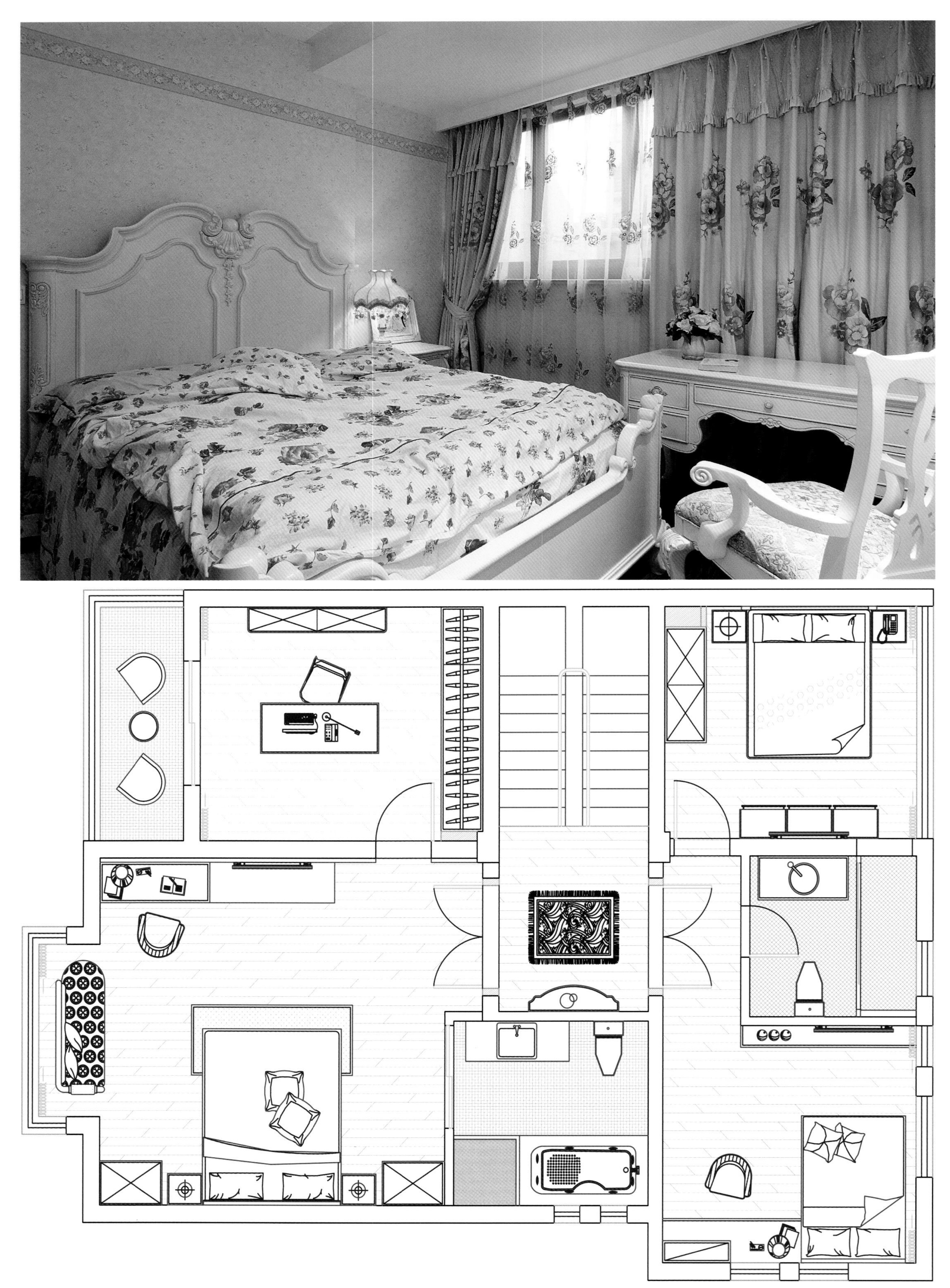

Plan　　平面图

本案的设计风格定位为美式风格，首先是考虑到业主的特殊的生活方式，业主的身份和社会地位，最重要的是体现业主对生活的热爱、对一些事物的怀旧情愫以及追求自由的生活态度和信念。

设计作品中运用到的材质和设计造型多数是将传统美式风格和居室本身所在地的地域文化的融合，通过这种独特的风格演变，将大宅得天独厚的气质和品位再次提升。在整个户型的优化和改造中，整个设计空间都以强调“空间进深感”为核心去重新塑造空间，深宅大院才能方显“大宅本色”。

“品茗长窗下，悠然以忘坐”。这就是此次美式风格大宅居室设计的感悟和真谛。

ZHONGSHAN TIAN YI
VILLA

中山天乙别墅

- Design Company: Scale Design Co., Ltd., Dongguan
- Location: Dongfeng Town, Chungshan
- Area: 284 m²

- 设计公司：东莞市尺度空间设计有限公司
- 项目地点：中山市东凤镇
- 项目面积：284 m²

European interior design which is fanned vary in social status is synonymous with luxury and sufficient splendor to boost sales in relevant building materials, furniture and light fixtures.

However, such prevalence leads to unprofessional treatment of details, the source of dilettante imitations of original fineness.

The proposed layout, however, is an exception. Simplified European style is expressed in dominant cream-white, whereas profound brownish black and warm-toned yellow aptly proclaim luxury. Unfettered by traditional extravagance, the design embodies relaxation, coziness, family love and unique aesthetic perception. Providing leisure reserved only for the aristocracy, the residence appeals to targeted clients, as every inch is especially refined for the dainty upper class.

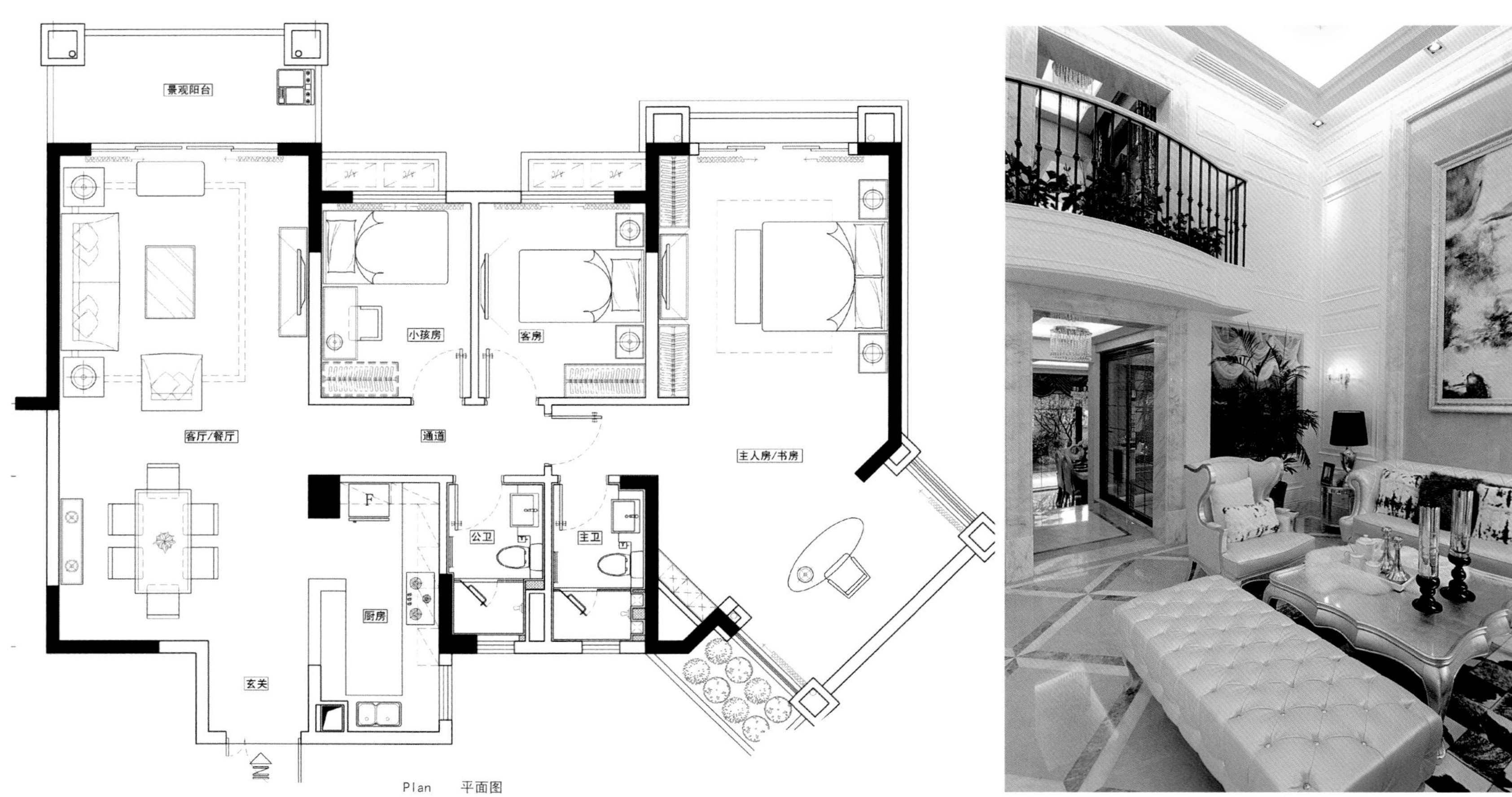

Plan　平面图

欧式风格在家居中最能体现出豪华感，这类风格在家装中普遍流行，得到不同身份业主的追捧，此现象可通过建材业、家具业、灯具业等了解其风格在市场上的热衷度。

而现在欧式风格在行业中出现泛滥，缺乏对细节的推敲，往往呈现的设计效果达不到国外经典的工艺及美观细节。

所以这次我们提出的简约欧式风格设计概念，采用暖色的米白作为室内的主色调，用沉稳的黑棕色、华丽高雅的暖黄色适宜的点缀出空间的奢华印象。年轻的色彩鄙弃了传统风格中的金碧辉煌，呈现居住者独特品位的同时，又使人感受到家的自然、舒适与温馨。通过对每个空间的详细推敲，设计总结出了高端居住群体的居家心理，模拟出一个休闲恣意的贵族家庭的生活情境，触动消费者的购买欲望。

ROAMING THE LAND OF EUROPA

漫步欧罗巴

- Designer: Li Haiming (LIHAIMING STUDIO)
- Location: Wanda Plaza
- Area: 168 m²

- 设计师：李海明（李海明工作室）
- 项目地点：万达广场
- 项目面积：168 m²

The designer: Every project turns a new page of excitement, filling my mind with brilliant inspirations. My talents reside in designing rather than lecturing. My design proposals are always welcomed surprises as well as my own pride.

设计师独白：每当一个新的设计开始时，犹如打开一张崭新的白纸，可以纵情挥洒的心态总是令人兴奋不已。只是实在不太善于用言语表述自己的思想，总是在设计方案出来后才会引起客户的惊喜，此刻，也是最有成就感的时候。

客卧室
GUEST ROOM
书房
STUDY ROOM
卫生间
TOILET
主卧室
OWNER BED ROOM
卫生间
TOILET
衣帽间
CLOCKROOM
次卧室
BED ROOM
阳台
BALCONY
厨房
KITCHEN
阳台
BALCONY
餐厅
DINING
客厅
LIVING

Plan　平面图

SIEMENS

TONGDE ELITE'S VILLA, KUNMING

昆明同德极少墅

- Designer: Zhang Zhiwei
- Location: Kunming, Yunnan
- Area: 220 m²

- 设计师：张植蔚
- 项目地点：云南昆明市
- 项目面积：220 m²

The residence is designed for a family of five (a couple with two children and an elderly), who share an appreciation for simplicity and elegance.

The hallway cabinet, the elder's wardrobe and the shoe cabinet are attentively hidden from view as an ingenious compromise of functions and aesthetics.

Original layout is optimized to provide additional floorage for every section, achieving a rhythmic perception of space. Spare passages are transformed to hold a piano.

Catering to the clients' taste, the atmosphere retains elegance despite of the residence's gorgeous components.

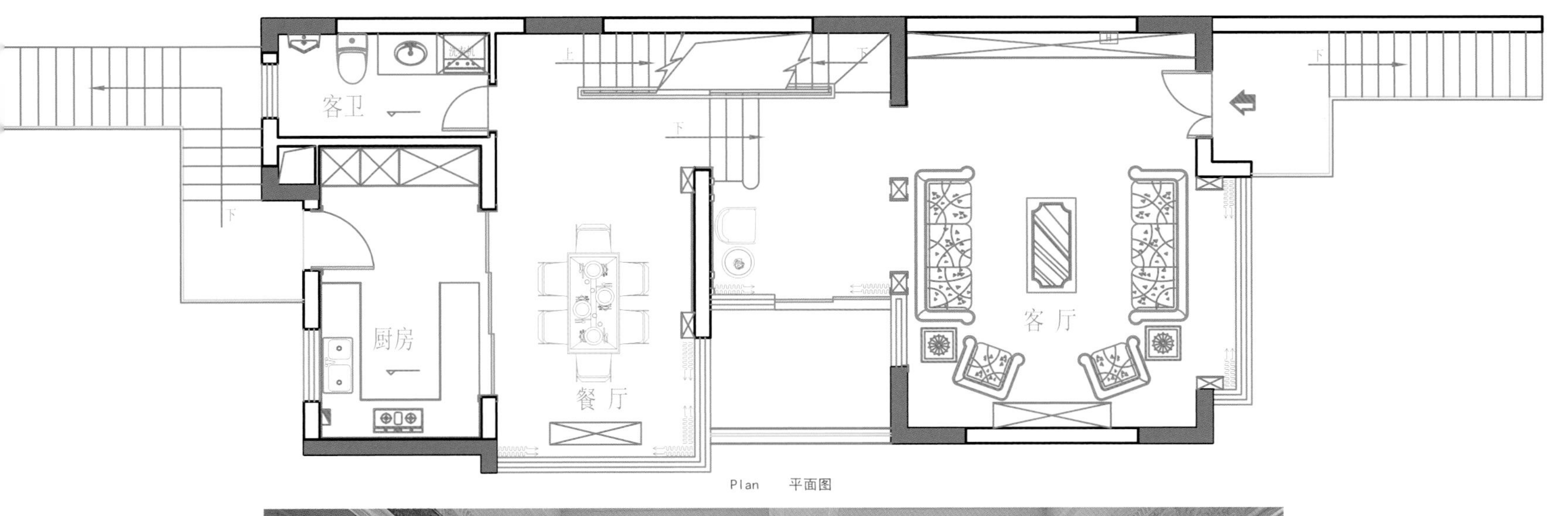

Plan　平面图

欧

业主为一家五口，男女主人、一儿一女、一位老人，整体都比较喜欢清新、雅致的家居氛围。

在整体设计空间布局时，考虑到将功能与美观联系起来，因此将门厅的功能柜、老人使用的储藏柜、鞋柜等功能性柜体（包括饮水机）全部安排至隐藏柜内，以达到美观的效果。

在空间布局中，优化了原有的空间，使得在进入每一个空间时，都有相对应的“缓冲区”，以增强节奏感，而原先一些利用不上的过道空间，也应用起来，做成钢琴区等区域。

在设计风格上，虽然采用了一些相对华丽的材质，但整体仍控制在一个比较典雅的氛围内，以吻合业主的要求。

IVY VILLA, CHONGQING

重庆常青藤

- Designer: Deng Dongchuan
- Location: Jiangbei District, Chongqing
- Area: 505 m^2

- 设计师：邓东川
- 项目地点：重庆市江北区
- 项目面积：505 m^2

Cigar bar, wine bar and home theatre, what else could a successful and cultured urbanite crave for? The garden resembles a locale in a pastorale, where loquats, mangoes and grapefruits flourish alongside with fragrant blossoms and greenery.

Greater lighting, ventilation and coziness are bonuses of a transformed exterior, allowing the scenery outside to pour in with least hindrance. Limpid water flows from the front to the back garden, a pleasant babbling disturbance to the villa's tranquility. This is a Chinese courtyard that both sings a European comception.

The exterior and interior, as well as different rooms are subtly coordinated by spatial transformation. Volume is visually maximized while separated functional areas are reasonably integrated.

The wooden suspended ceiling is a small repository of details while the living room joist qualifies as the designer's touch of genius, exemplifying the villa's innovative ornamentation.

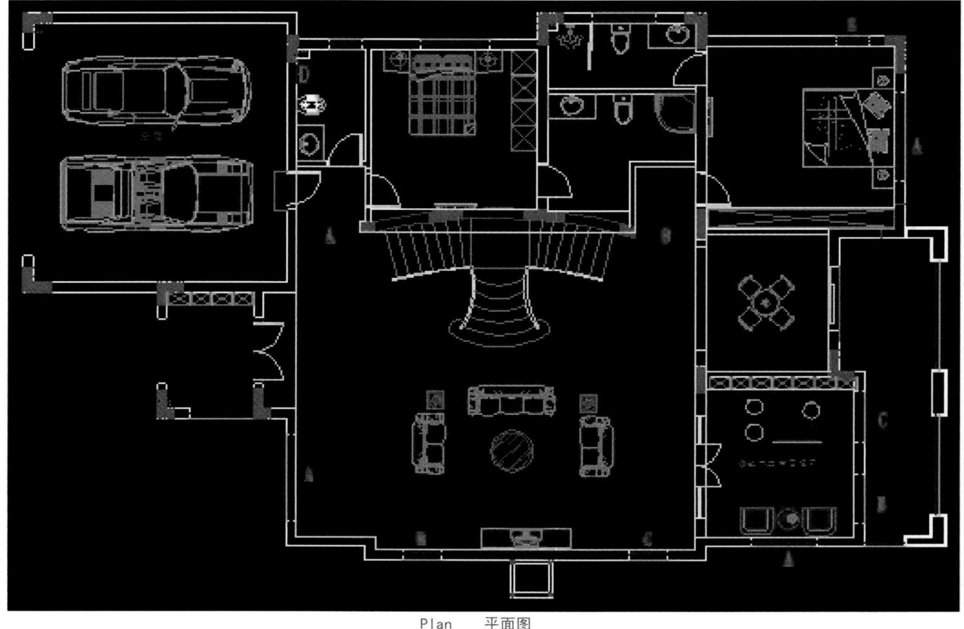

Plan　平面图

雪茄吧，红酒房及视听室满足了都市成功一族对生活品质的追求，庭院的绿化除了观叶观花植物以外还种植了如枇杷、芒果、柚子等果树让业主享受田园般的生活。

此外，作者还对别墅的外观进行了一些改造，目的在于尽量把户外的绿色引入室内，增加采光与通风，增加居住的舒适性。外环境上利用水来连通前后花园，也通过流水来划分动静的区域，中式的格局，欧式的意境。

设计师在整个空间改造上主要采用通透的格局，不光是室内和室外的通透，还有室内各个空间的通透，不同的区域有分隔有联系，达到视觉的最大化，空间的合理化。

设计师在设计选材上也创新点：整个木质吊顶的处理让细节比较丰富，客厅托梁处理比较满意。

THE SHINING PEARL OF YANGTZE , CHONGQING I

重庆江上明珠 I

Designer: Sun Yuan' an

Location: Chongqing

设计师：孙元安

项目地点：重庆

The dreamy and deluxe interior design provides the urbanites in the fast lane with a poetic elegance. The dominating peacock blue and subdued gold qualify as refined modern hues, while luxurious walnut floor extends to the entire residence. Adorned with leather, velvet, kidskin and mirrors, the residence's innate luxury pervades.

梦幻而又奢华的家居设计给生活在繁华都市中的人们带来一种朦胧但又雅致的生活方式。设计选用锈金色和孔雀蓝作为主色调，表现了一种现代的雅致风格。奢华的胡桃木地板使整个室内空间融为一体，而皮革、天鹅绒、小山羊皮和镜子的采用更是体现了奢华的本质。

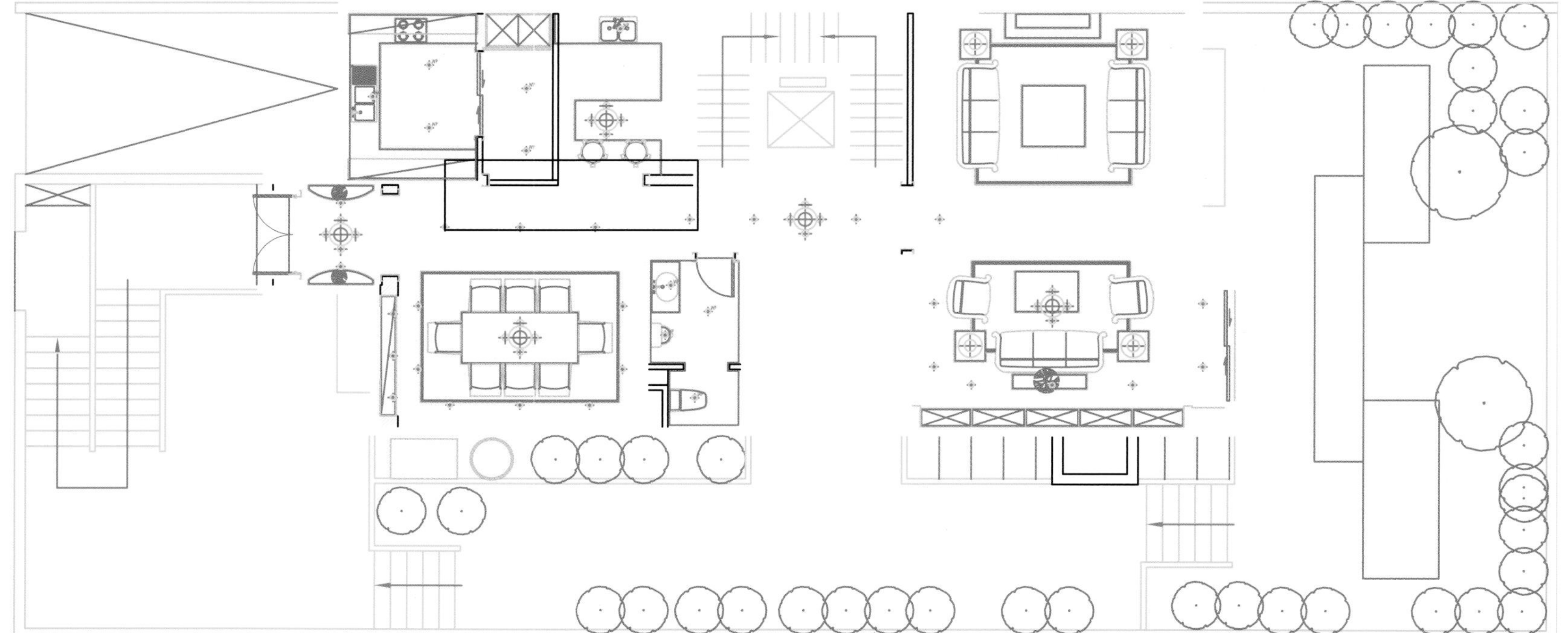

Plan　　平面图

OPERATION
VALKYRIE

THE NATURE AND PROPERTIES
OF ENGINEERING MATERIALS
VLSI ARRAY PROCESSORS

Statistics Explained

THE SHINING PEARL OF YANGTZE Ⅱ, CHONGQING

重庆江上明珠Ⅱ

- Designer: Sun Yuan' an
- Location: Chongqing

- 设计师：孙元安
- 项目地点：重庆

This case designed in Neoclassic style. The walls are decorated mainly in white, for that the bright color can enlarge and light up the space. The high-back sofa, wool-made material of the dinning chair and the simple style of the decoration, all these make the space full of fashionable atmosphere and also show us the graceful feeling of classic style. The back-shape ceiling in master bedroom is unique and individual characteristic. With the warm colors, the whole space create a cozy and romantic atmosphere. The faint yellow wallpaper give us a relaxed and easygoing home feelings.

本案的设计风格为新古典主义风格。墙面以白色为主要格调，亮眼的色彩令空间尽显宽敞明亮。高背的时尚沙发，简约的餐桌椅绒质的面料以及简单的制作，令居室空间萌生现代时尚气质，但不失古典风格的气韵。主卧室的背脊型吊顶，个性十足。整个居室空间以暖色调来营造卧室环境的氛围，有一种温馨浪漫的气息油然而生。淡黄色的墙纸，带给人一种恬然、悠闲的家居感受。

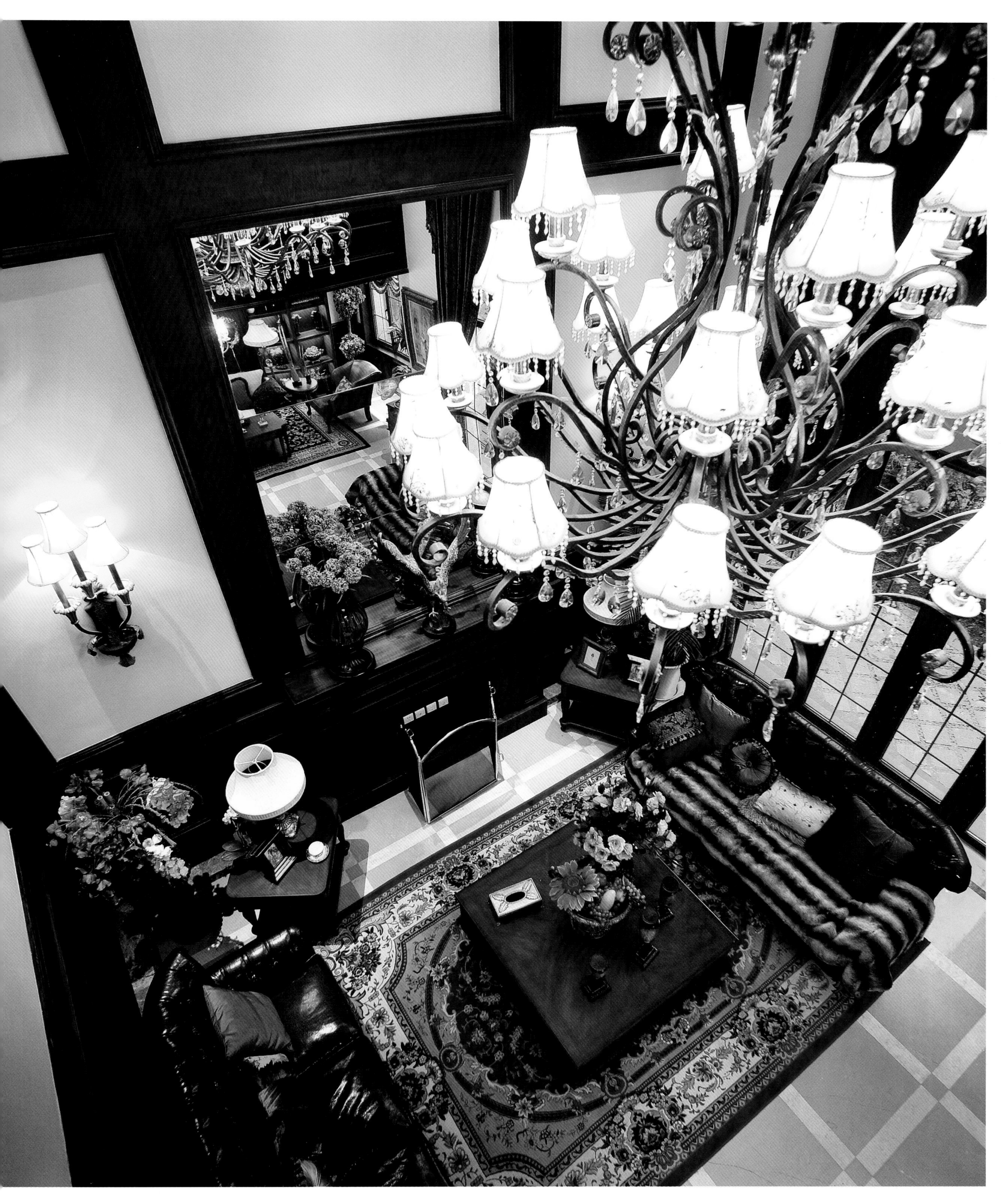

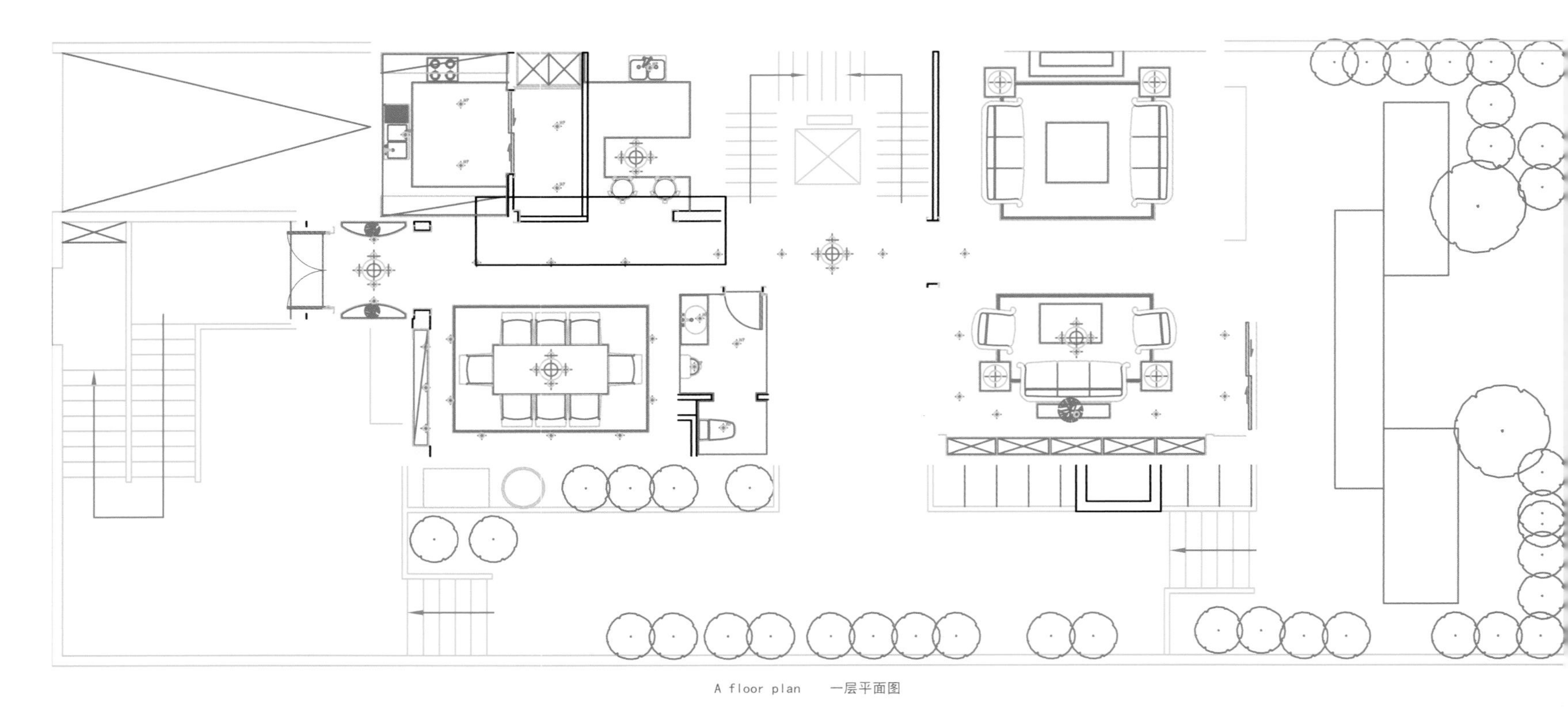

A floor plan　　一层平面图

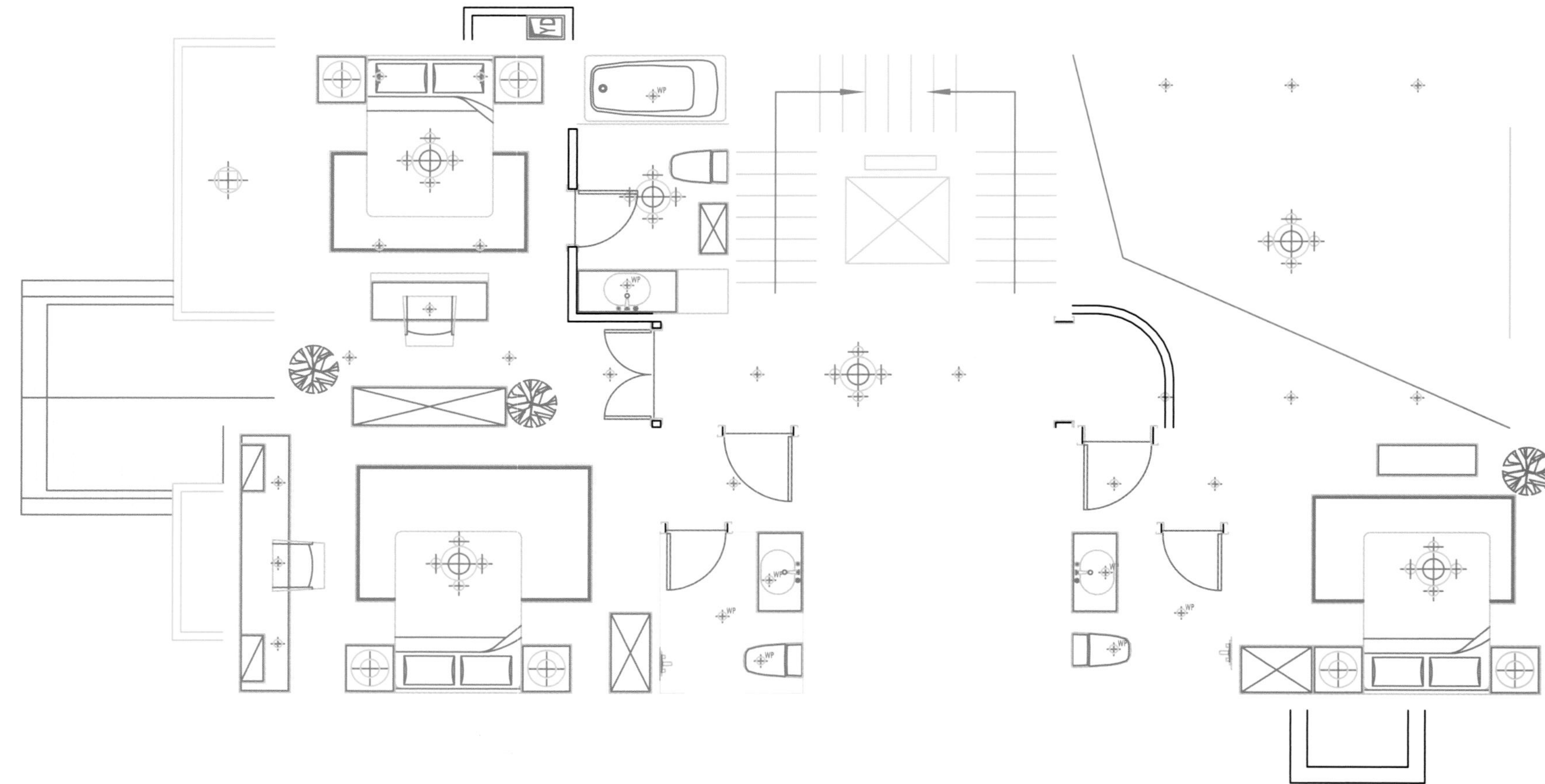

The two floor plan　二层平面图

Georges RICHARD

URBAN ERA T1B SAMPLE HOUSE

城市时代 T1B 样板房

- Designer: Xu Shuren (Shenzhen Di Kai Interior Design Co., Ltd.)
- Location: Huizhou, Guangdong
- Area: 134 m²

- 设计师：徐树仁(深圳市帝凯室内设计有限公司)
- 项目地点：广东惠州
- 项目面积：134 m²

A Pygmalion imagination inspires the design motif. Therefore, the residence, is the incarnation of a perfect lady, whose poise is graceful and charm breathtaking. Deluxe but not extravagant, or proud, the residence reserves elegantly simplified classic lines. It is a place where comfort and relaxation pervade. Diamond-velvet fixtures resemble the lady' finest gown, while the pearly white varnish is her tender skin. Sophisticatedly deployed mirrors extend her luxurious air, whereas elaborate furniture and ornaments exemplify her inner beauty, providing refined embellishment that guarantees coziness and tranquility. The lady's flowery scent can be easily imagined, which lingering an amazement to all.

XGNCP
XGNCP
XGNCP
Entertainment

XGNCP
XGNCP

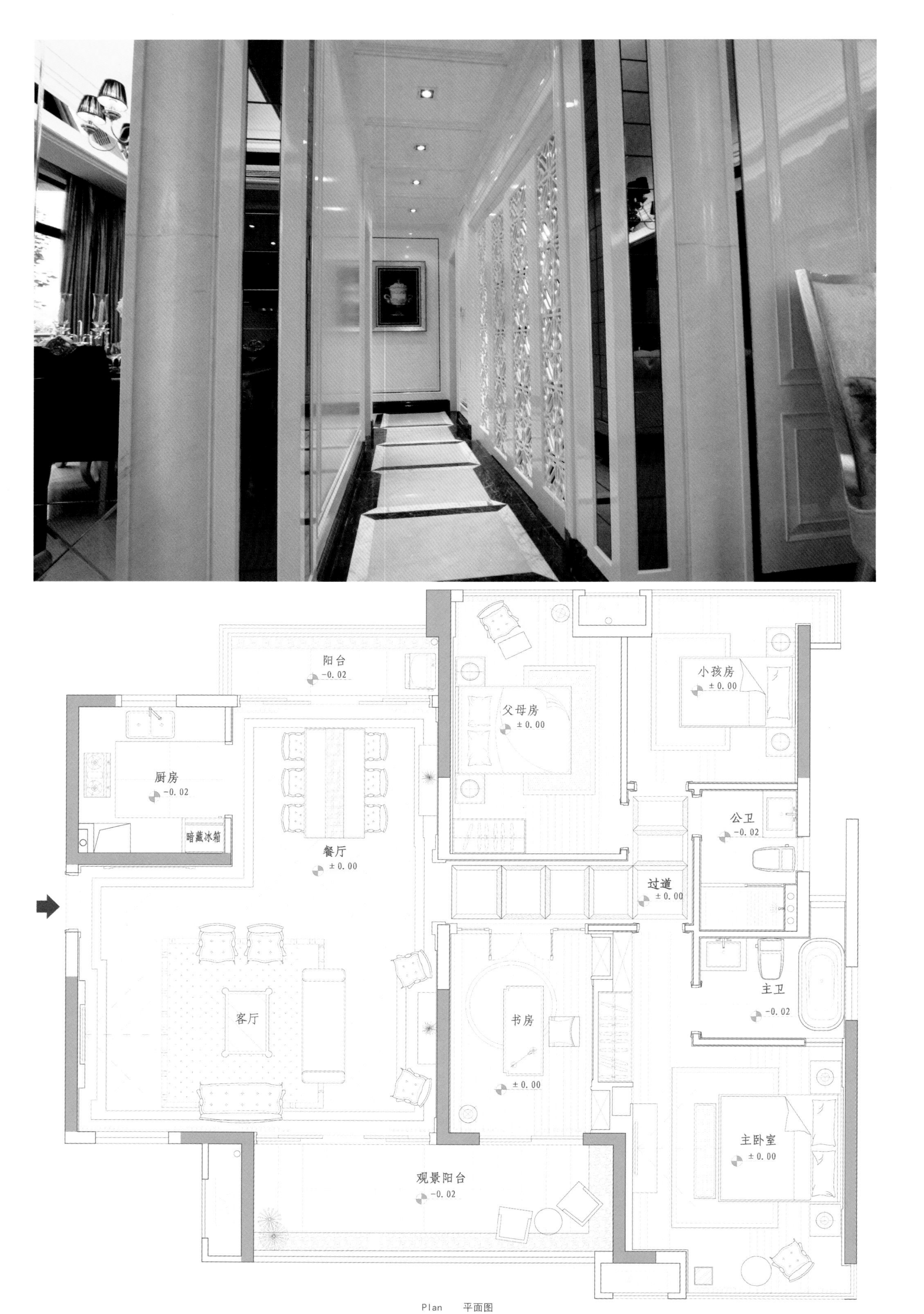
阳台
-0.02
小孩房
±0.00
父母房
±0.00
厨房
-0.02
暗藏冰箱
餐厅
±0.00
公卫
-0.02
过道
±0.00
客厅
书房
±0.00
主卫
-0.02
主卧室
±0.00
观景阳台
-0.02

Plan 平面图

A
XGNCP
A
XGNCP

案设计概念是基于具有高贵气质与慑人风采的女性形象的想象，风格奢华而不奢靡，贵气而不张扬，简化的古典线条，带着一种悠闲舒适感。空间采用的钻石绒硬包以及镜面对空间的延伸，让空间的质感细腻地呈现出了别样的奢华感。大面积的珠光白漆饰面增加了空间的温婉之气，极富心思的家具配饰，隐约地显露了它的内在美。设计师想让样板空间能让人感觉娴静舒适，让它高贵优雅，倾城倾国之气弥散开来。

ZHONGGUAN FINE VILLAS

中贯美墅馆

- Designer: Li Jianhua
- Location: Changshu, Jiangsu
- Area: 400 m²

- 设计师：李建华
- 项目地点：江苏常熟
- 项目面积：400 m²

The concise and fashionable Neoclassicism frees the project from complicated traditional details.

Heydays of Peace (Jia He Sheng Shi) residential quarters target at the middle and upper class who crave for a cosy and relaxing family life after a tedious day of work. Significant considerations are given to the layout. The resulting project integrates commodiousness of public places and privacy of bedrooms. The living room, extending right down to the balcony, is granted abundant grace and space, where sunshine pours in freely. Therefore, the barrier between guests and householders is removed, creating a place for geniality and ease.

Black, pale coffee, silver and white dominate the residence, avoiding excessive contrasts in shades. Chaste wallpapers also provide fresh simplicity and elegance.

An open dining room is the best choice for building a strong family. Fine food and life can be savored, with views to the sunlit living room and a small garden hidden in groves.

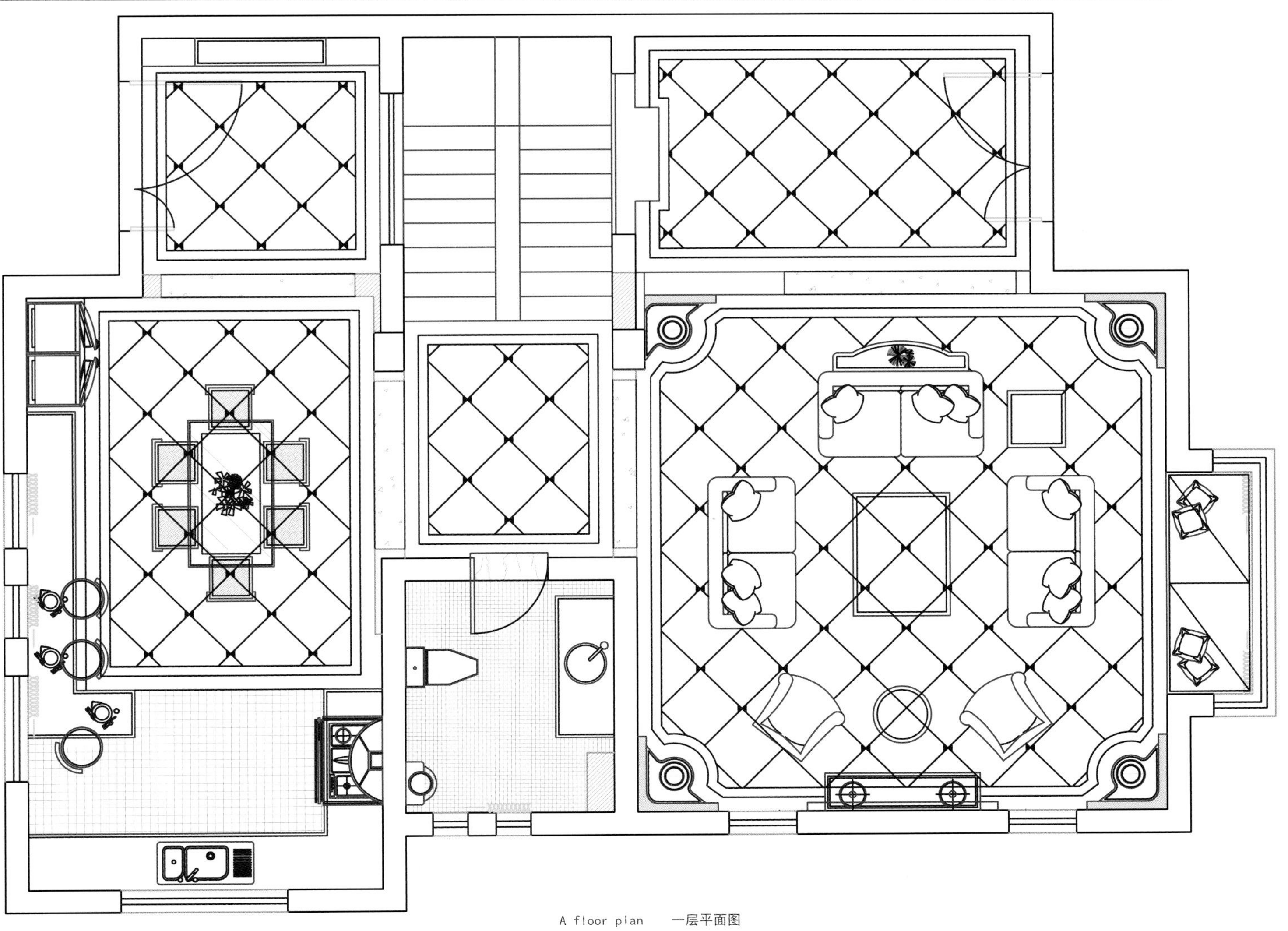

A floor plan　一层平面图

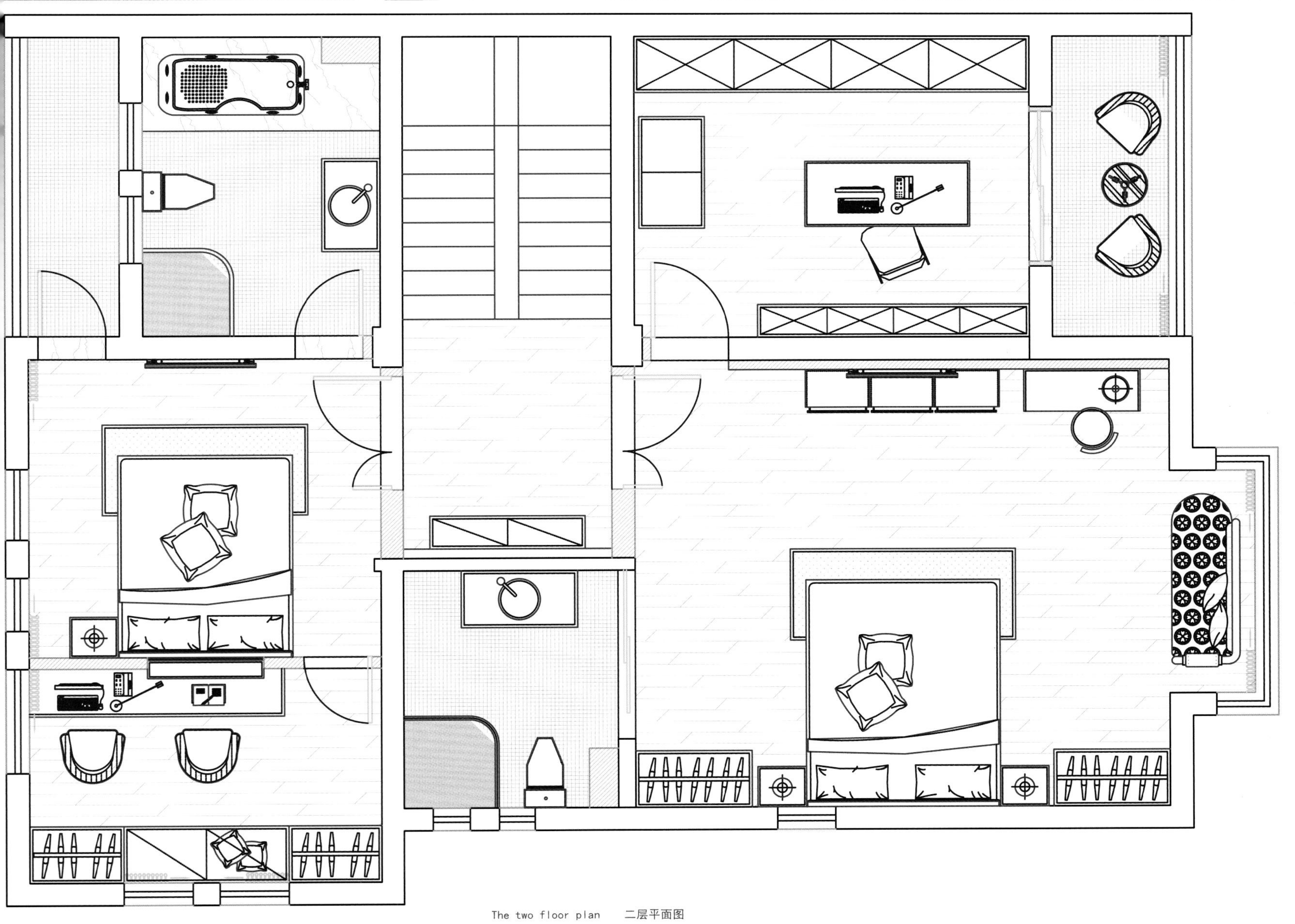

The two floor plan　二层平面图

本案摒弃了繁琐的传统古典风格，运用简约与时尚的手法构成本案的新古典主义设计风格。

“家和盛世”的目标客户群是中高端的白领家庭，他们在工作之余，追求舒适自然的家庭生活。设计师在室内功能划分及布局设计上，倾注于创造公共场所的大开放性与卧室私密性。因此，客厅设计就要求开阔而且大气。本案中，客厅与阳台相连通，阳光可以自由地进入，增强了室内的活力，使家庭成员与访客之间不受约束，自由且随意。

本案以白色、黑色、咖啡色、银色四种同色系为主要色调，壁纸的应用只是略做修饰，并未有炫目壁纸的过分装点，所以还保持着空间整体的清新淡雅，使之具有高雅的氛围。

倾注于家庭成员的交流沟通，开放式餐厅无疑是最好的选择。一边毗邻幸福满溢的客厅，一边毗邻绿意盎然的小花园，让你品味美食之时也能品味生活。

DUPLEX SAMPLE HOUSE

样板房复式

- Designer: Jason Poon (J&B Design Studio)
- Location: Shenzhen
- Area: 84 m^2

- 设计师：潘冬东（J&B设计事务所）
- 项目地点：深圳
- 项目面积：84 m^2

Situated in Da Dong Cheng Residential Quarters II, Pingshan New Area, Longgang District, the 84-square-meter sample duplex is aristocratic and elegant, embellished with light shades in great proportions and occasional black and gold. Exotic amberina glassware, mirrors and shiny black steel adorn the drawing room while a unique sideboard adds to the glamour.

Glittering ornaments reign in the entire residence with limpidity and dazzle, as if they are stars visible night and day, contributing to the charisma and vigor of a home. Bright colors, well-chosen materials and refined styles in the nursery, study and master bedroom perfectly exemplify the householder's upper-class status while additional ornaments and details provide the luxurious European residence with abundant family geniality.

本案为龙岗坪山新区大东城二期的样板房，面积为 84 平方米的复式户型，整个空间中采用了黑色，金色与大面积的浅色作搭配，彰显了高贵典雅的气质。客厅中异型艺术玻璃的使用，镜面和黑钢的结合，特色小酒柜的处理都成为了亮点所在。

在本案中，作者用最闪亮的装饰包裹家的全身，不论在白天还是夜晚，家从此展现自身的迷人魅力。装饰品剔透的质感和夺目的外观为家注入了无限活力。在主人房与儿童房中强烈地体现了卧室主人的身份特征。品质格调的主人房，书房无论是在色泽与材质，细节与装饰，都让整个设计在欧式的华丽中无线延伸出家的乐趣。

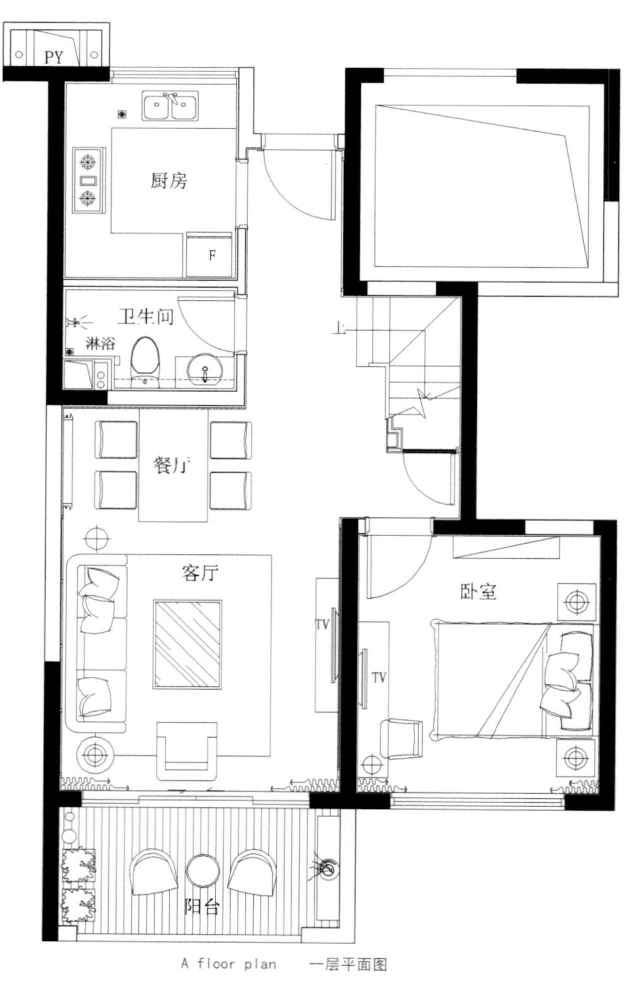

A floor plan　一层平面图

Windows 7

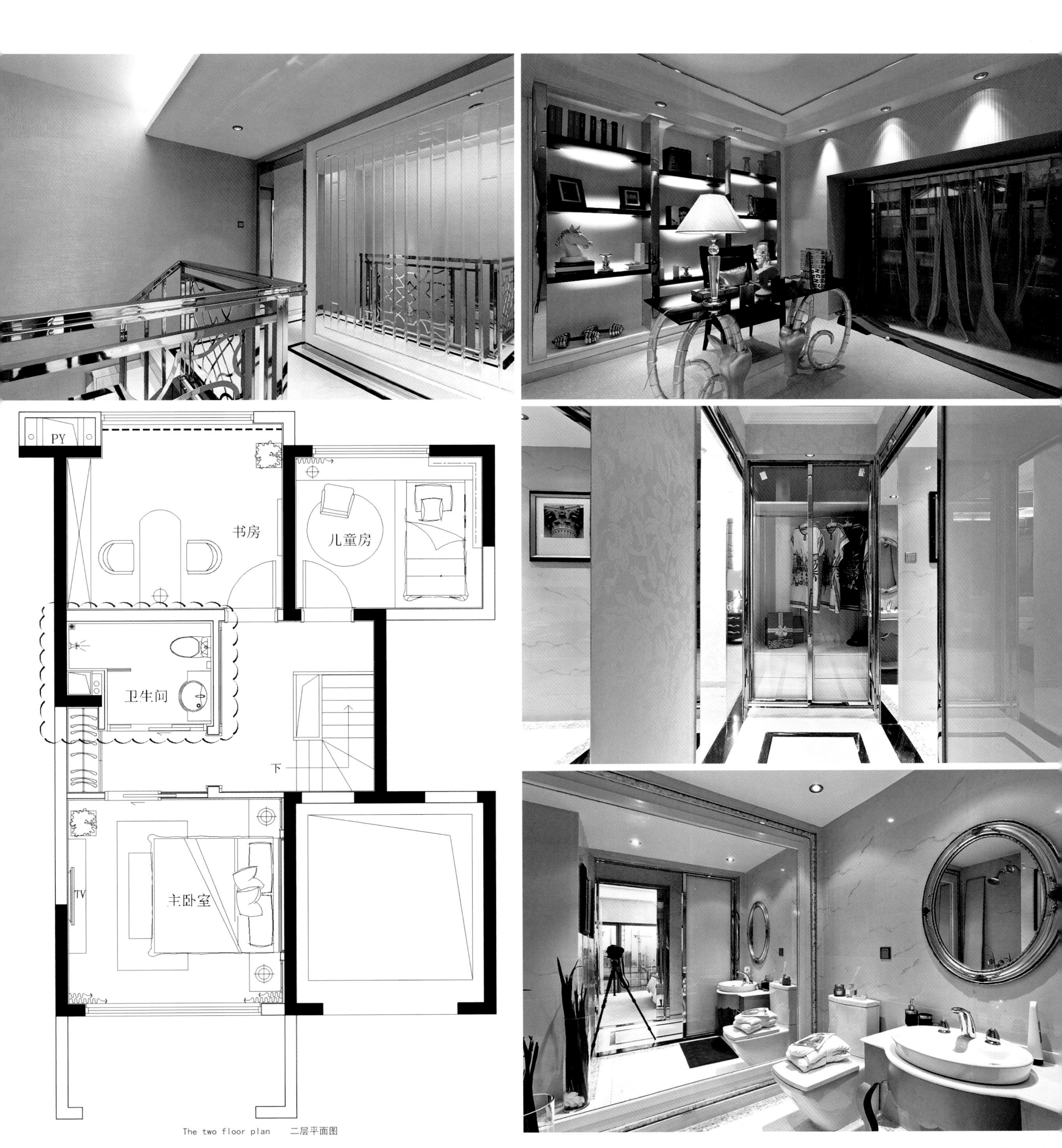

The two floor plan　二层平面图

VILLA ARCADIA, CHONGQING

重庆阿卡迪亚别墅

- Designer: Hu Wenbo
- Location: Chongqing
- Area: 1200 m²

- 设计师：胡文波
- 项目地点：重庆
- 项目面积：1200 m²

The layout flourishes with classic elements are targeting at a client gifted with sober appreciation of luxury and life. Embellished with a bold palette, the residence enjoys an overwhelming result, when its commodiousness provides an instant guarantee to relaxation. Elements ranging in hue and texture indicate the householder's identity, status and values expressing disapproval of excessive details. In conclusion, the designer intends aesthetics of this project, and that of life, to reach all hearts, so that is the essence of a masterpiece.

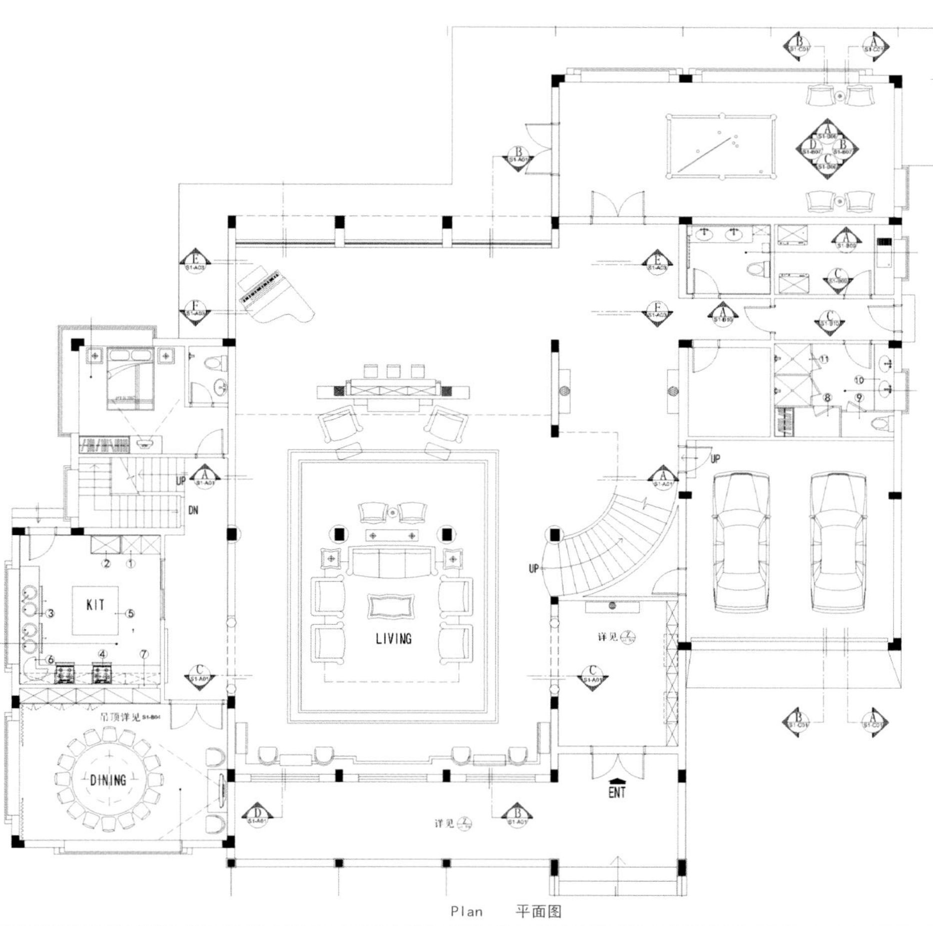

Plan 平面图

本案运用了古典的设计元素，市场定位是奢华，让业主身处其中可以充分享受生活。作者对于此案环境风格的采用非常地大胆、果断，同时本案的空间颜色搭配使得整体效果非常有气势，非常的大气。此外，开阔大气的空间布局也让人神清气爽，非常的舒适。设计师不赞同过于奢华繁琐的设计，觉得那样会让人感到压抑，同时设计师觉得无论是从设计之美，还是生活之美，都应该让之成为人人都能拥有的东西。因为好的设计是服务于大众的。

NORA ROBERTS

"TO MY BELOVED" SAMPLE HOUSE, CHONGQING

重庆心语样板房

- Designer: Sun Yuan' an
- Location: Chongqing

- 设计师：孙元安
- 项目地点：重庆

The fashionable European project is dominantly embellished with unique Mediterranean blue, echoing a life by the Big Blue. Modern decorative pictures add a natural and lovely atmosphere. In the meantime black dining table, blue and white chairs, amberina glasswares, transparent candelabrums and carpets and jars varying in size provide a sweet locale for romance.

时尚欧式：整个空间中以地中海独有的蓝色为主色调，营造出一个海边生活的居家场景。黑色的餐桌配以蓝白色的餐椅、地毯、透明的玻璃杯、烛台和大小不同的罐子，仿佛在述说着甜蜜的爱情；墙上相框作为装饰画，体现出现代都市感，无不彰显着自然、温馨的生活氛围。

Hers

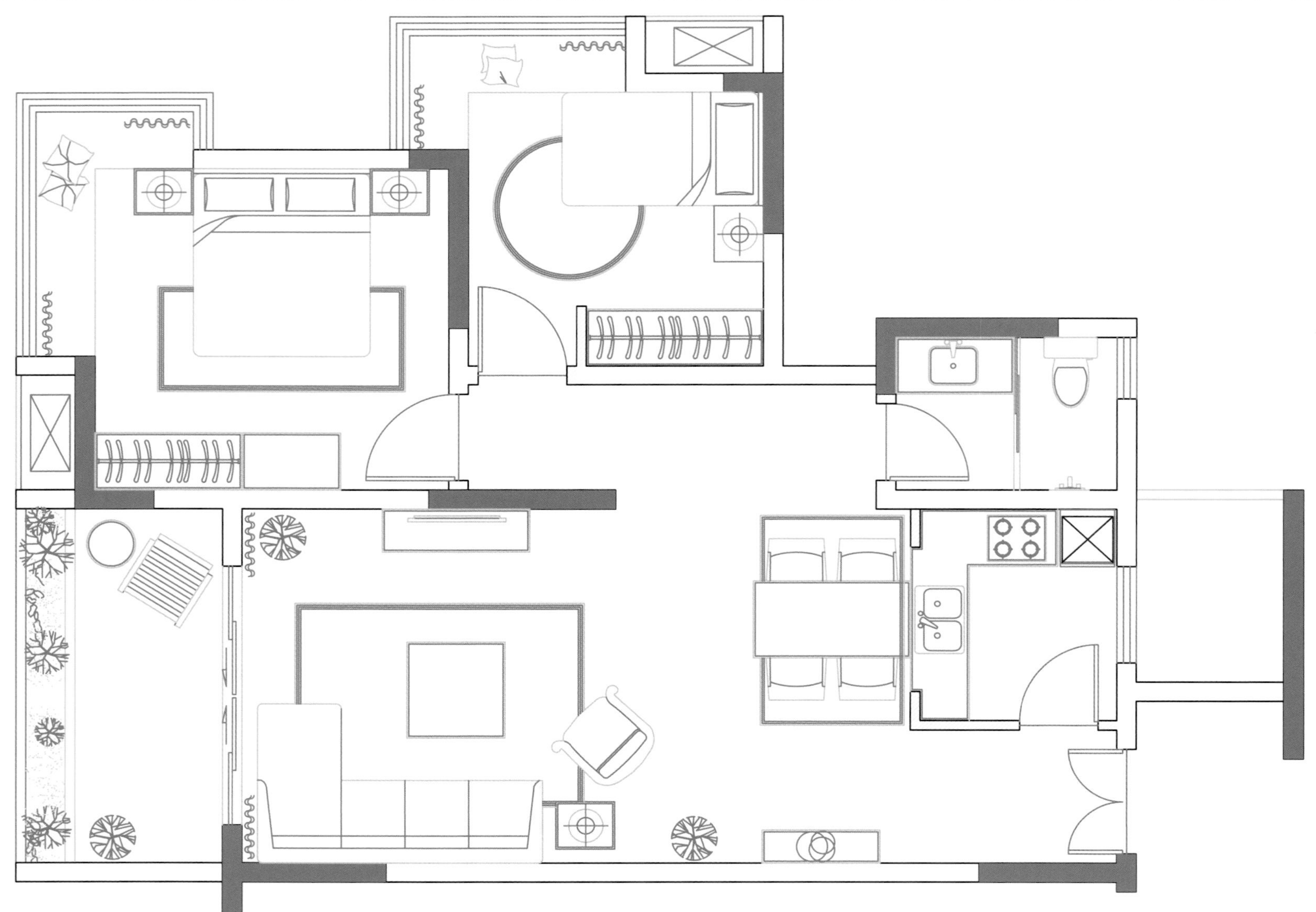

Plan　平面图

NEOCLASSICAL SAMPLE TOWNHOUSE, HEFEI

合肥联排别墅新古典样板房

- Designer: Xie Ruixue
- Location: Hefei, Anhui
- Area: 420 m²

- 设计师：谢瑞雪
- 项目地点：安徽合肥
- 项目面积：420 m²

The residence's neoclassical interior is consonant with its exterior, while a garden subtly interlinked indoor and outdoor space.

Layout is improved by transforming the original patio and providing additional indoor floorage while contributing to spatial gradation, enjoyment and financial value.

Materials are well chosen. European wallpapers and Chinese silk ones with delicate handmade patterns are ingeniously integrated with BISAZZA mosaic in the hallway betters overall ornamentation.

本案建筑设计风格为新古典主义风格，作为本案的小环境因素，室内设计与建筑保持着一致性，花园内景观作为室内和室外的过渡，是一个软性结合点。

在布局上，作者利用了原建筑的天井，改造成为室内可利用的空间，增加使用面积及空间利用价值，使得空间更具有层次感及趣味性。

在设计选材上，作者采用丝绸手绘壁纸与欧式墙纸结合，混搭了一些中式的元素进去。门厅采用 BISAZZA 的马赛克拼图，增加了空间的装饰效果。

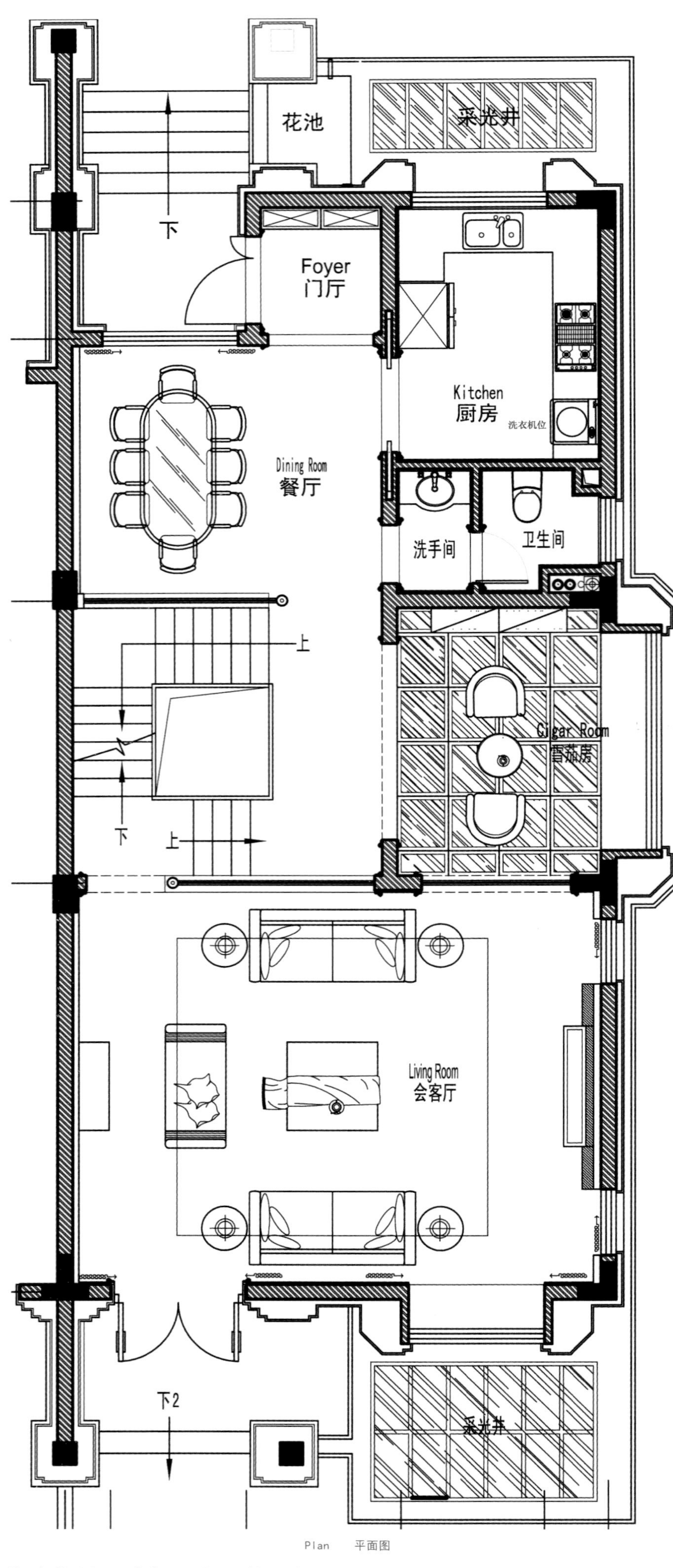
花池
采光井
下
Foyer
门厅
Kitchen
厨房
洗衣机位
Dining Room
餐厅
洗手间
卫生间
上
Cigar Room
雪茄房
下
上
Living Room
会客厅
下2
采光井

Plan 平面图

ANHUI HUA DI KUNGKUAN

安徽华地公馆

- Designer: Wang Qinjian; Wang Hao (Shenzhen Mok Environmental Art Design Co., Ltd)
- Location: Anhui
- Area: 127 m²

- 设计师：王勤俭；王浩（深圳市墨客环境艺术设计有限公司）
- 项目地点：安徽
- 项目面积：127 m²

The design motif embraces a luxurious yet low-profile lifestyle, where a sober European refinement reigns unfettered by gaudiness. The delicate living room and garden provide perfect locales to savour life and romance with the fragrant coffee and an engaging novel.

The layout is aptly transformed to subtly-arranged dimensions. The new doorway and a fine cabinet sit on opposite ends of the hallway, joined by a living room on the left and a piano room on the right. The dining room recedes into an atrium. Frosted mirrors and glass, together with the opposite rooms by the doorway, visually enlarge the deluxe place. Another atrium is turned into a guestroom, guaranteeing the privacy in the grand main room.

The residence combines luxury and modern romance, streamlines and straight lines, delicate furniture and unique wallpapers and a palette of bright shades. It may even rival a palace. Restoring the purest romance, the designer impresses his target clients with a chaste and deluxe innovation.

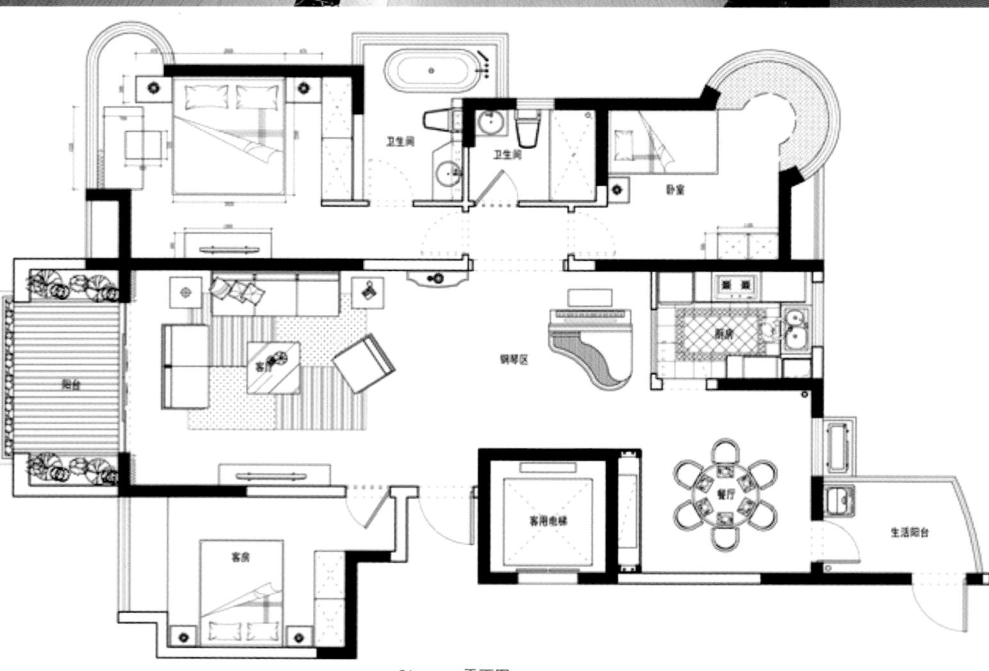

Plan　平面图

样板房想传递一种低调奢华的生活态度，既有欧式生活的贵气与精致，又无金碧辉煌的浮华，无论坐在客厅还是花园小憩一杯浓浓的咖啡，阅读一本小说，品味一种生活，一种浪漫……

合理的空间格局改造使得空间序列完整。经过调整后开门正对玄关案台，左侧客厅、右侧钢琴厅（把餐厅调到挑空花园空间位置）；布局双厅概念，再结合磨花玻璃、磨花镜虚拟反射，使得空间奢华大气。利用另一挑空花园做一间客房，保证了大宅的主客私密性。

整套大宅的设计是奢华主义与现代浪漫主义的完美结合，流线形与直线相互搭配，现代奢华的家具精美华丽，明亮的色彩相互交错，所有豪华的家饰，个性的墙纸相互呼应，此情此景充斥着浪漫的诱惑力与华贵感，而表现出王者风范。设计者试图在设定的消费族群中，将一种华丽贵气的表达形成另一套符号系统而深植人心。

IMPERIAL CROWN

IMPERIAL CROWN

STYLISH URBANITE RESIDENTIAL QUARTERS

风格城事

- Designer: Ge Xiaobiao (Jin Yuan Men Co., Ltd.)
- Location: Ningbo city, Zhejiang
- Area: 150 m²
- Photographer: Liu Ying

- 设计师：葛晓彪（金元门设计公司）
- 项目地点：浙江 宁波市
- 项目面积：150 m²
- 摄影：刘鹰

The neoclassical design motif, inspired by consistent, splendid and intriguing ornamentation of Catania (a port city of Sicily, southern Italy), is finalized after thorough consultations with the householder, whose preferences and requirements are considerately attended to. Integrating the classic and modern, as well as the functional and artistic, Italian architecture embodies grace, vogue, aristocracy and a fine aesthetic taste. Highlighting all these traits, the residence is pervaded by Italian style.

The designer also revives traditional handwork and materials, presenting a romantic Catania work with chaste shades, displays and embellishments.

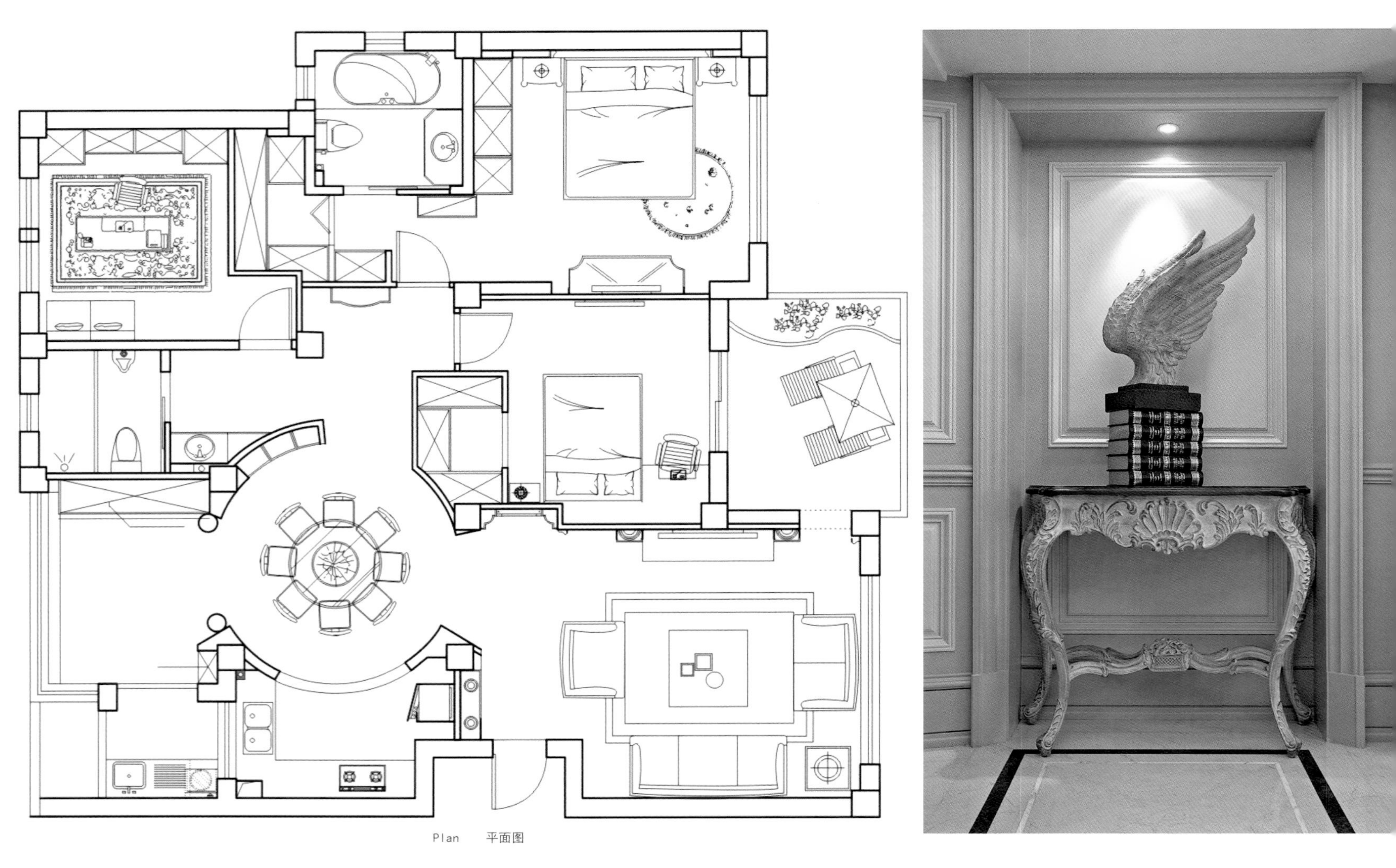

本案位于鄞州区。经过和屋主的多次沟通之后，设计师了解到屋主的喜好和需求，最后将此案的设计风格定位为新古典，并且选择了以意大利南部西西里岛的卡塔尼亚名城为灵感之源。从美丽的卡塔尼亚城，设计师发现其装饰风格和谐，耀眼并且充满诱惑，此外，意大利风格的建筑是古典和现代、功能与艺术的结合体，拥有艺术生活品位，简约，但是却不失时尚与高贵！因此设计师致力将此案打造布置成一个生活和艺术相融合，同时也混合着简约之风的意大利风情之家！

在此案中，设计师使用传统的材料和手工工艺，用简约的造型和色彩，配上陈设，打造出富有浪漫情调的卡塔尼亚之恋。

Harness Racing

DEFINING POST-MODERN LUXURY

后现代的奢华诠释

- Designer: Liao Xinyao
- Location: Masterland Residential Quarters, Nanjing
- Area: 420 m²

- 设计师：廖昕曜
- 项目地点：南京玛斯兰德小区
- 项目面积：420 m²

Unique ornamentation best exemplifies the designer's mastery of style, exquisite carving frees any piece of furniture from duplication, well-chosen prickets add a romantic atmosphere whereas custom-made decorative paintings in every corner further the effect.

Cream-colored walls, comfortably padded sofas and crystal chandeliers provide the residence with luxury and romance, perfected by lines in European style extending from carpets to the ceiling. Stylish neoclassical furniture, elaborate decorations and prickets are attentively coordinated in the roomy and well-lit hall, enriching the otherwise monotonous locale.

The kitchen and dining room owe their visual spaciousness, brightness and perception of depth to grand French windows, cream-colored and specially shaped upper beams, as well as light-shaded walls. Refreshing scenery pours in unobstructed from these windows, soothing an urbanite's surfeit of the bustling city. Removable decorations are subtly arranged, integrating the residence while allowing for new visual impacts at every step.

设计师对这套别墅风格的拿捏，最显著的是对其特有鲜明设计风格的配饰运用。每一件家具的雕花都精致典雅，但样式绝不雷同；各式各样的蜡台让整个空间增加了几许情调；随处可见的订制的装饰画，则渲染了独有的设计氛围。

从米黄色墙面、软包沙发、地毯到天花板上的欧式线条及水晶吊灯，让整个空间弥漫着奢华与浪漫的气氛。整个大厅设计得通透、明亮，别具风格的新古典家具的摆设及精致惟妙的饰品、烛台的搭配，使空间不再单调，使生活更富有情趣。

厨房与餐厅空间独立大块的落地窗结合米黄色顶梁造型和浅色墙面使空间开敞而明亮，极富层次感。宽大的落地窗，使户内户外景观相呼应，打造出现代人追求人与自然的和谐共处的生活品位之氛围。错落有致的软装饰将整体建筑空间串联起来，呈现出移步异景的意境。

The first floor plan　一层平面图

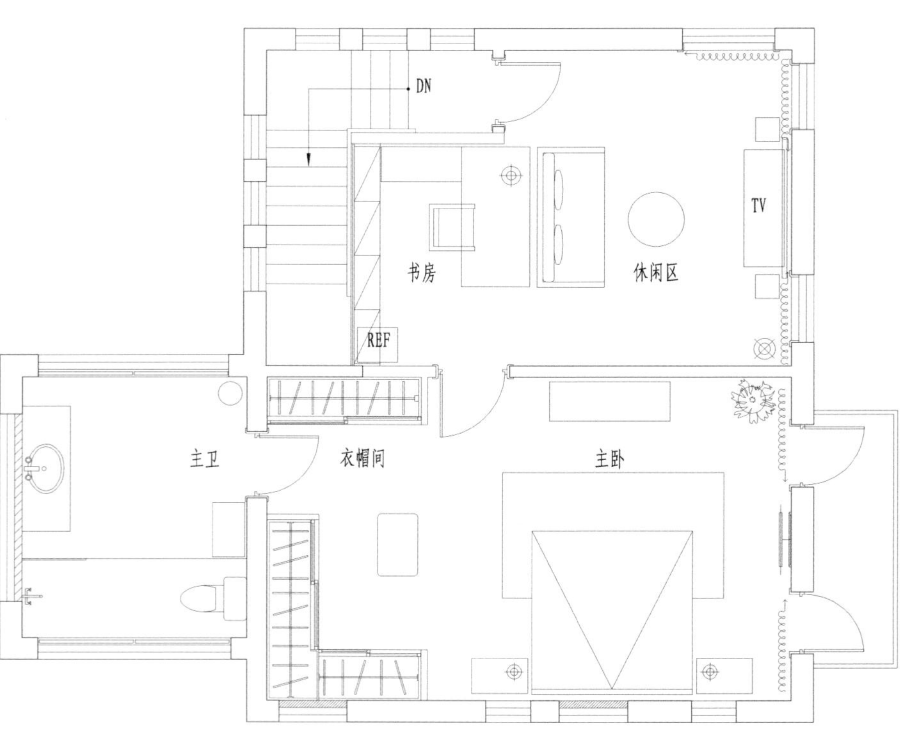

The three floor plan　三层平面图

HOT SPRING VALLEY

温泉山谷

- Designer: Huang Lirong (Zhong Ce Decoration, Kunming)
- Location: Kunming
- Area: 220 m²

- 设计师：黄丽蓉（昆明中策装饰（集团）有限公司）
- 项目地点：昆明
- 项目面积：220 m²

The essence of luxury resides in low profile and modesty. Modern-Aristocracy Classic style, finalized for the design motif, is the craving shared by the cultured new-rich. Believing that showy vogue, pretended dignity and superficial ostentation can never define prosperity, they demonstrate a fine aesthetic appreciation in every detail.

A sophisticated designer bears in mind that a residence is the extension of its owner's personality and the style should naturally reflect the householder's taste. Therefore, none of the project's spatial arrangements, materials, furniture and ornaments is deployed for the sake of style. Carefully attending to the client's sober refinement and palate for fashion, the residence provides a stage in the depths of glory to free the true self.

低调才是真正的奢华，设计之初给这套房子新贵古典的风格定位。什么是新贵古典？也许可以说是现代新贵们内心深处的一个情结——时尚、高贵，炫富时代已经成为了过去，真正的富裕不用炫耀，那是每一个角落、每一个细节暗藏着的精致体现。

空间、材料、家具、配饰这一切不是为风格服务，而是为房主的品位服务，这才是设计的本质，用什么名词形容主人，就该用什么名词形容他的房子。我的客户是时尚、雍容、大气，所以这套房子也该如此。居住在灿烂深处，延续生活秀场，随心所欲地展示自我。

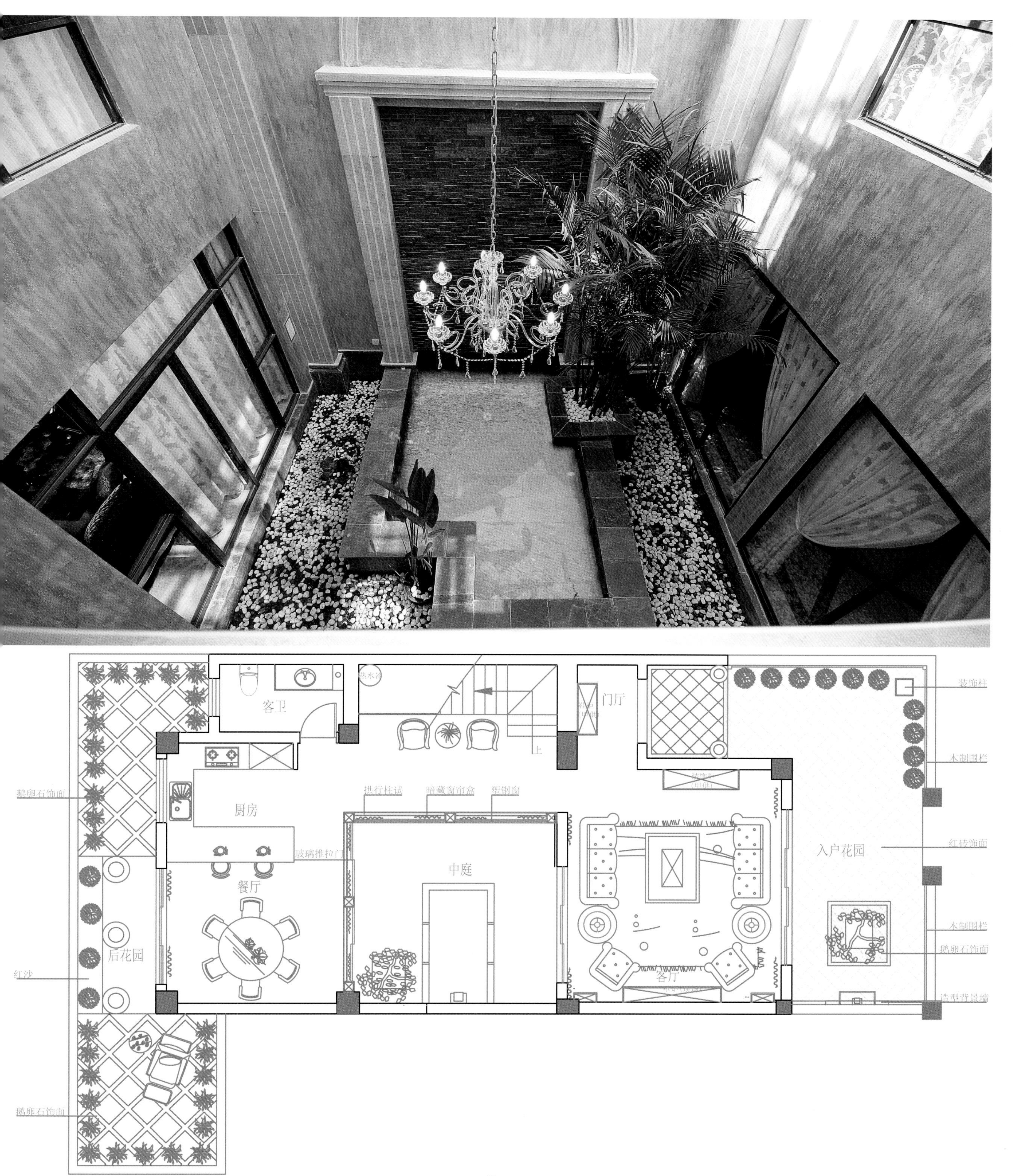
客卫
热水器
上
门厅
装饰柱
厨房
拱行柱试
暗藏窗帘盒
塑钢窗
鹅卵石饰面
木制围栏
红砖饰面
玻璃推拉门
中庭
入户花园
餐厅
后花园
红沙
木制围栏
鹅卵石饰面
客厅
造型背景墙
鹅卵石饰面

Plan　平面图

请勿使用!

VANKE TANGYUE TOWNHOUSE

万科棠樾联排别墅

- Designer: Han Song (Horizon Space, Shenzhen)
- Location: Tangyue, Dongguan
- Area: 311 m²
- Photographer: Deng Xuebin

- 设计师：韩松（深圳昊泽空间设计有限公司）
- 项目地点：东莞棠樾
- 项目面积：311 m²
- 摄影：邓雪彬

The project enjoys unbeatable location. Situated in the town of Tangyue with natural lakes and vegetated slopes for company, it satisfies the dream shared by many: a return to nature guaranteed by easy transportation. Tailored to the middle and upper class, the design motif is finalized as horse. The residence provides improved enjoyment and quality of life, as reflected in the elegantly simple European style, graceful curves, perception of depth and furniture of low-profile luxury. Grace, dignity, relaxation, passion and confidence pervade the entire place.

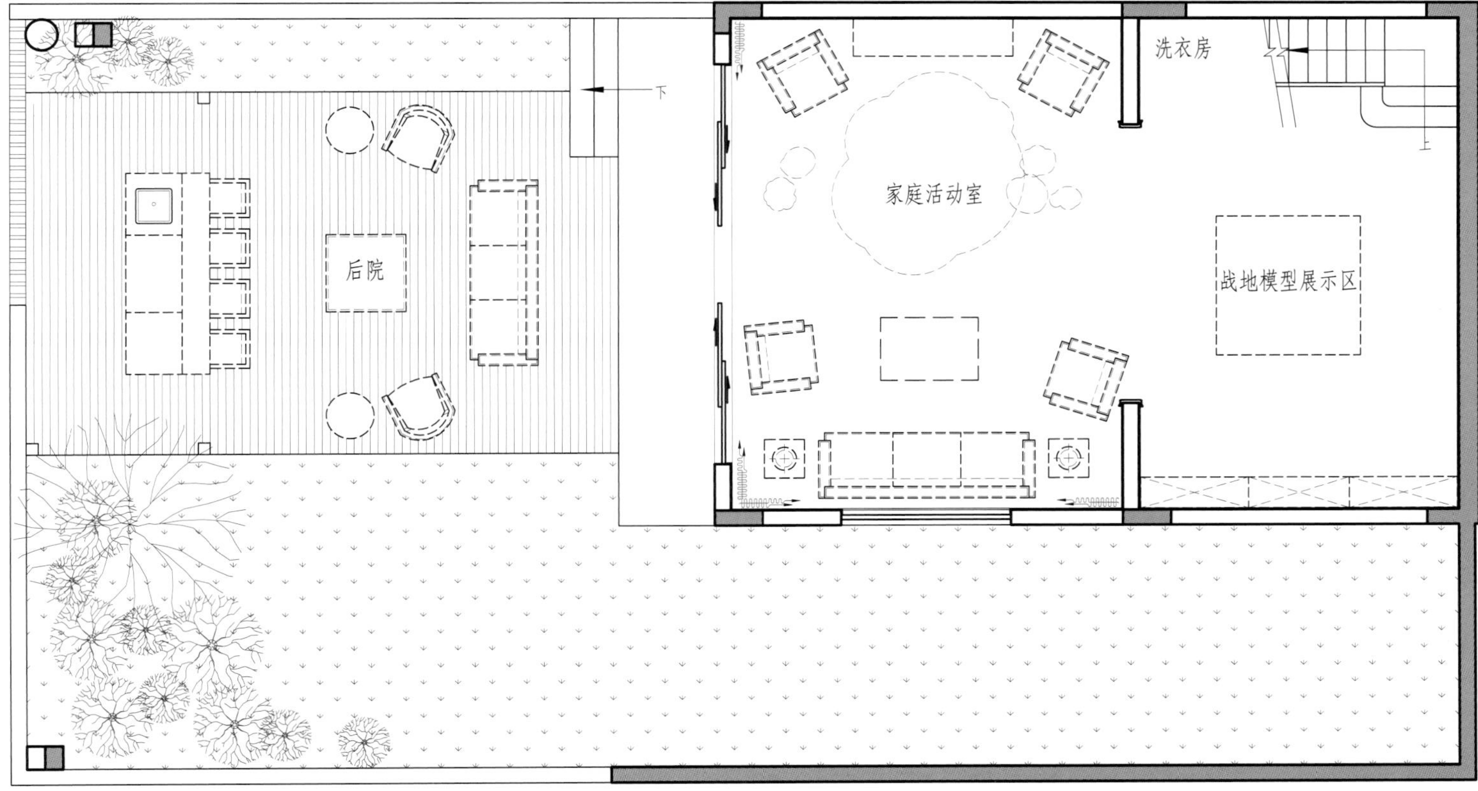

Basement plan　地下室平面图

本案位于东莞棠樾，坐落于天然湖泊与自然山景之中。既满足人们亲近自然，实现别墅梦想的愿望，又有比较方便的交通条件。作品定位于希望提升生活质量，享受生活的中高端客户群。因此在设计中运用了优雅舒适的简约欧式的风格，层层连贯的空间、柔和的线条，低调中略带奢华的家具，以及贯穿于所有空间中“马”的主题，共同营造出一种雍容、放松，同时又充满激情自信的人生态度．

A floor plan 一层平面图

KELLY HOPPEN HOME

BOOKS TO INSPIRE
KELLY HOPPEN HOME

VANKE EASTERN SHORE

万科东海岸

- Designer: Wang Wuping
- Location: Shenzhen, China
- Area: 800 m^2

- 设计师：王五平
- 项目地点：中国深圳
- 项目面积：800 m^2

The beaded curtain by the dining room, glittering and limpid, is the incarnation of gentle raindrops, pleasantly slowing down an urbanite's pace. The chandelier, consonant with the aristocratic neoclassical chairs, is reflected in an elegantly simple black dining table of glass, while the unique sideboards add to the room's splendor.

The honeycomb ceiling of the piano hall provides unawares an artistic atmosphere while allowing light through, visually enlarging the place. Similar visual effect contributes to the partition behind the sofas. This fretted wall of bold geometric patterns divides and enriches the huge drawing room while providing a perception of depths.

The residence boasts subtle spatial arrangements. The top-floor terrace is transformed into a miniature driving range, while the tiered garden level is embellished with lawns, trails, garden chairs and a pergola, promising a locale for coziness and leisure.

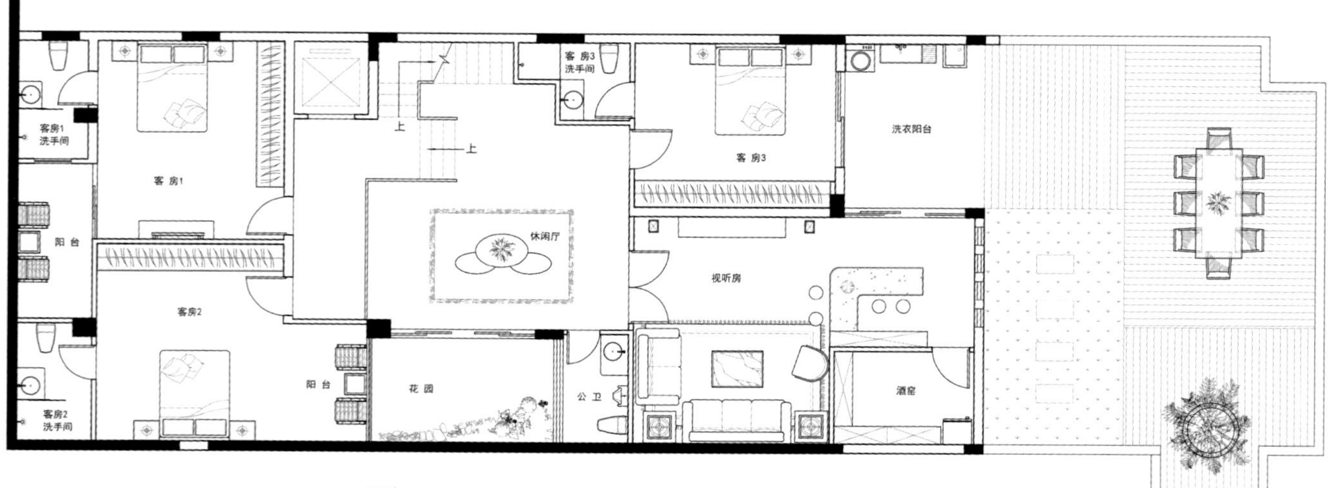

地下室平面图　Basement plan

餐厅旁边的水线帘，水珠顺着钢线慢慢地滑下，颗颗晶莹剔透，滑动的是一种生活节奏。黑色玻璃的大餐台，没有诉说太多的繁锁，倒是旁边一点新古典风格的餐椅，透露着高贵的气质，与餐厅的吊灯相得益彰。旁边的别致餐柜，也让餐厅多了一份光鲜。

钢琴厅的蜂巢天花，不经意间就给这里增添了不少艺术氛围。因为透光，拉伸了钢琴厅的视觉高度。沙发背后的几何形体与蜂巢天花异曲同工，传递着空间的尺度。因为客厅之大，才有了设计师勾勒出这个大气的几何形体，让空间不再单一地重复着简单。

景观设计是本案一大亮点，顶层屋面露台设计改造成一个高尔夫练习杆，负一层花园设计也层次叠出，草坪、步道、护外木架、休闲椅等，这些不无营造出一个舒适，怡人的景观环境。

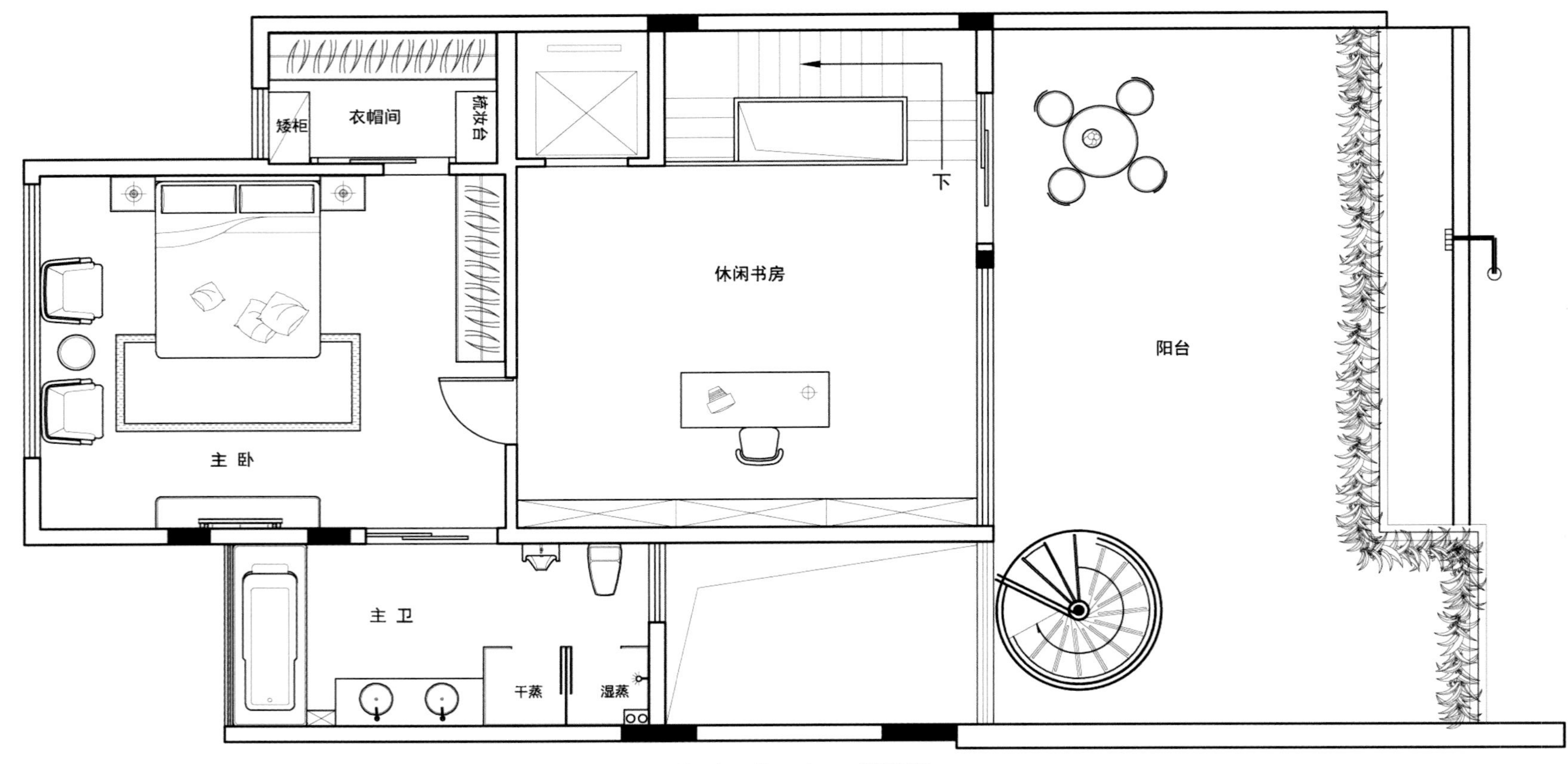

The three floor plan 三层平面图

PORTMAN HOUSE

波特曼建业里

- Designer: Wubin
- Location: Shanghai
- Area: 408 m²

- 设计师：吴滨
- 项目地点：上海
- 项目面积：408 m²

Portman house is one project on stone gate building of the old Shanghai. The whole style of design is from old Shanghai in China. And the whole design style of lobby is like a living room and makes clients feel warm at home. In the middle of lobby, the designers arrange the sofas. At the back of the lobby, the projector equipment shows the alterative Chinese paintings. This creation of design makes the static space be more active. The designers make use of the vivid comparison between black and white. The ceiling is decorated with the white mirror frame. And the hang ceiling retains the natural wood color of the building. This idea shows the historical feeling in old Shanghai.

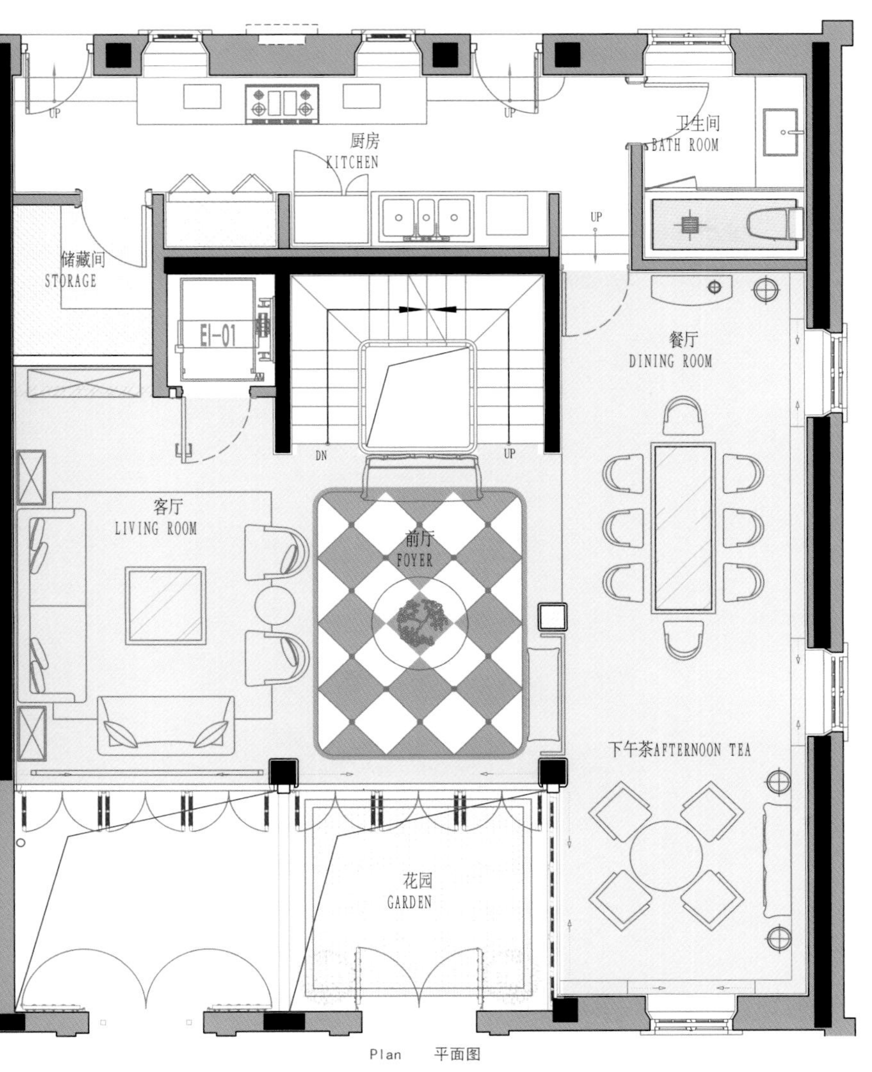
卫生间
BATH ROOM
厨房
KITCHEN
储藏间
STORAGE
E-01
餐厅
DINING ROOM
客厅
LIVING ROOM
前厅
FOYER
下午茶AFTERNOON TEA
花园
GARDEN
UP
DN

Plan 平面图

Portman house 是一个关于老上海石库门建筑改造的项目，因此整体设计风格彰显中式老上海风格。大堂整体设计为一个客厅的布局形式，为传递一种宾至如归的感受。在客厅的中央，布置了可供休憩的沙发。在背景处，通过投影设备播放着充满无限变化且具有中国意境的图案，使整个空间静中有动。

客厅的设计中运用了鲜明的黑白对比手法，顶棚将白色的空镜框作为修饰，吊顶则保留上海老建筑原有的自然原木色，无疑是对老上海沧桑历史感的一种诠释与传承。

YUANZHONG FENGHUA

远中风华

- Design Company: Sherwood Design Group (Huang Shuheng, Ouyang Yi, Chen Yijun)
- Location: Shanghai
- Area: 267 m²
- Photographer: Wang Jishou

- 设计公司：玄武设计（黄书恒、欧阳毅、陈怡君）
- 项目地点：上海
- 项目面积：267 m²
- 摄影：王基守

The project, one of the four flats in its building, enjoys a generous floorage. This urban castle, exclusively reserved for the upper class, owes its impressive vogue to intriguing statuettes and bold contrasting shades poured out in great proportions.

The residence fully demonstrates Art Deco.

The interior design highlights style revival, fusion, and innovation, integrating classic western fineness and an oriental perception of Renaissance Revival. The project, therefore, acquires European romance alongside with oriental grace, completed by a mixture of classic and modern decorations.

Charisma of old Shanghai extends the project's cultural deposits, offering a bonus to the cultured minds.

LIGHTART
Sherwood
Sherwood
New Style of spaces
New Style of spaces

□户为同栋四户中较大面积者，为了打造富豪门第、都会城堡的格局，因此大量运用对比强烈的色彩，□眼夸张的艺术造型显其不凡，以展现时尚庄园豪邸的大气尺度。

□rt Deco 装饰主义的精神在此大户中展现无遗。

□空间装修上，玄武设计运用西方古典工艺的严谨精湛工法，将东方新文艺复兴的精神注入其中。整体空间氛围传递着西方的浪漫，却也轻诉着东方的曼妙——透过古典与现代装饰艺术的交会融合，显示出复古、融会、创新与再生的精神。

老上海的清新魅力，为空间注入丰硕的生命力，也为居住者带来全新的心灵悸动。

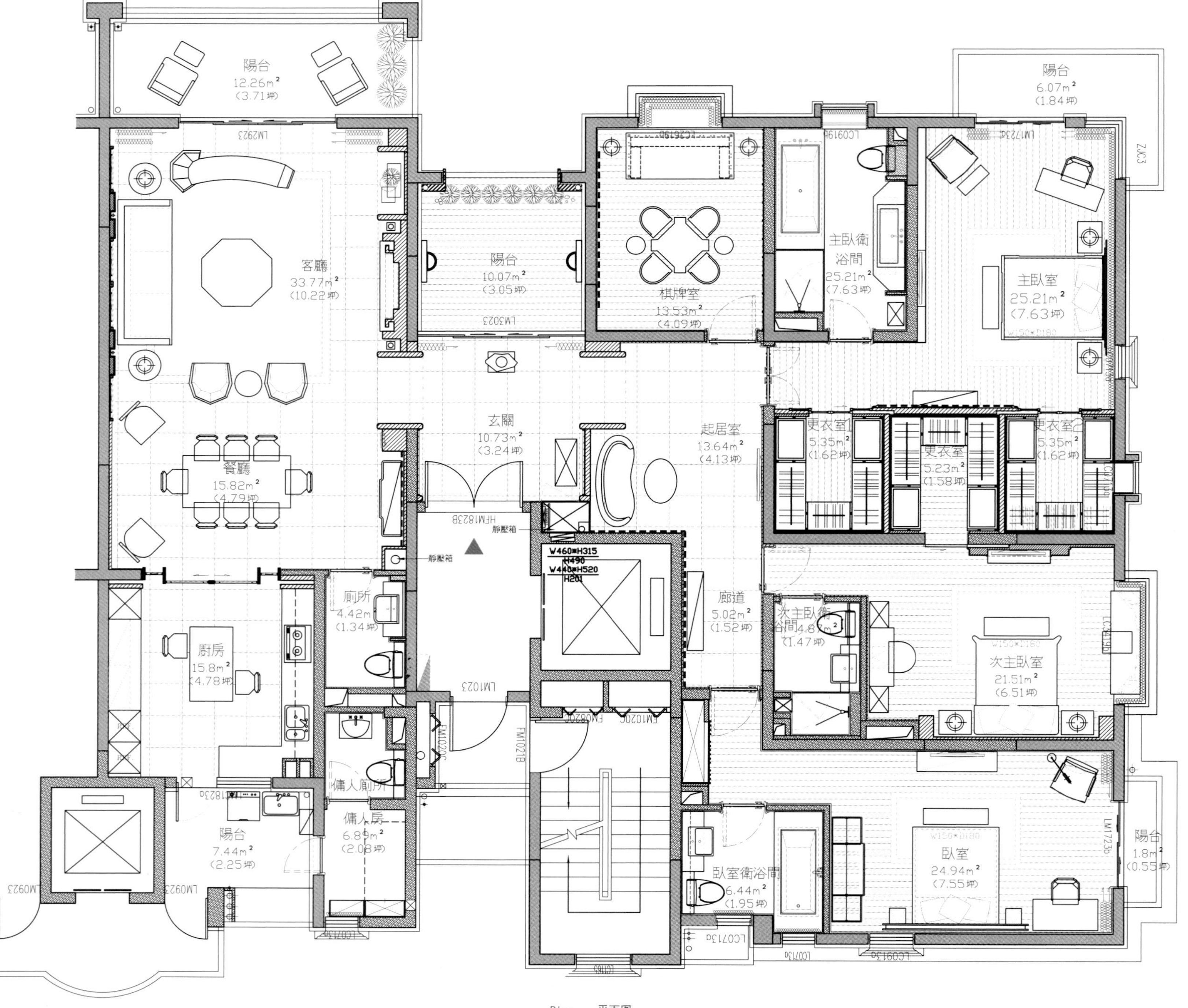
陽台
12.26m²
(3.71坪)
客廳
33.77m²
(10.22坪)
餐廳
15.82m²
(4.79坪)
陽台
10.07m²
(3.05坪)
棋牌室
13.53m²
(4.09坪)
主臥衛
浴間
25.21m²
(7.63坪)
陽台
6.07m²
(1.84坪)
主臥室
25.21m²
(7.63坪)
玄關
10.73m²
(3.24坪)
起居室
13.64m²
(4.13坪)
更衣室1
5.35m²
(1.62坪)
更衣室
5.23m²
(1.58坪)
更衣室2
5.35m²
(1.62坪)
廁所
4.42m²
(1.34坪)
廚房
15.8m²
(4.78坪)
廊道
5.02m²
(1.52坪)
次主臥衛浴間
4.87m²
(1.47坪)
次主臥室
21.51m²
(6.51坪)
傭人廁所
傭人房
6.89m²
(2.08坪)
陽台
7.44m²
(2.25坪)
臥室衛浴間
6.44m²
(1.95坪)
臥室
24.94m²
(7.55坪)
陽台
1.8m²
(0.55坪)
靜壓箱
W460*H315
H490
W440*H520
H201

Plan　平面图

FARGLORY SAMPLE HOUSE

远雄样品屋

- Design Company: Sherwood Design Group (Ouyang Yi, Chen Yijun, Lin Jiayun, Cai Mingxian)
- Location: Xinbei
- Area: 100 m^2
- Photographer: Wang Jishou

- 设计公司：玄武设计群（欧阳毅、陈怡君、林佳澐、蔡明宪）
- 项目地点：新北市新庄区
- 项目面积：100 m^2
- 摄影：王基守

The project demonstrates a surreal musical centered on vanity of life. Mingled illusion and reality challenge the viewer's wildest imagination, while fierce contrast adds to dramatic visual impact. Lines, totems, ornaments and furniture create dynamic variations and a perception of depth, inviting the viewer to join the fantastic show.

Baroque, a style of exuberance, grandeur and tension, conquered 17th century Europe, where pursuit of territory and wealth dominated all hearts, where science progressed despite of incessant wars. The turmoil intensified emotions, as reflected in Baroque's complicated details and flowing lines. Mastering its essence, Sherwood Design Group selects a subtle palette of black, white and grey. Further embellished with lines and ornaments of gold and silver, the residence is perfected by reflections in crystals, glass and shiny surfaces, resulting in dramatic visual contrasts and dynamics that redefine Baroque aesthetics.

MODERN GLAMOUR
MODERN GLAMOUR
Sherwood Stairs
LIGHTART

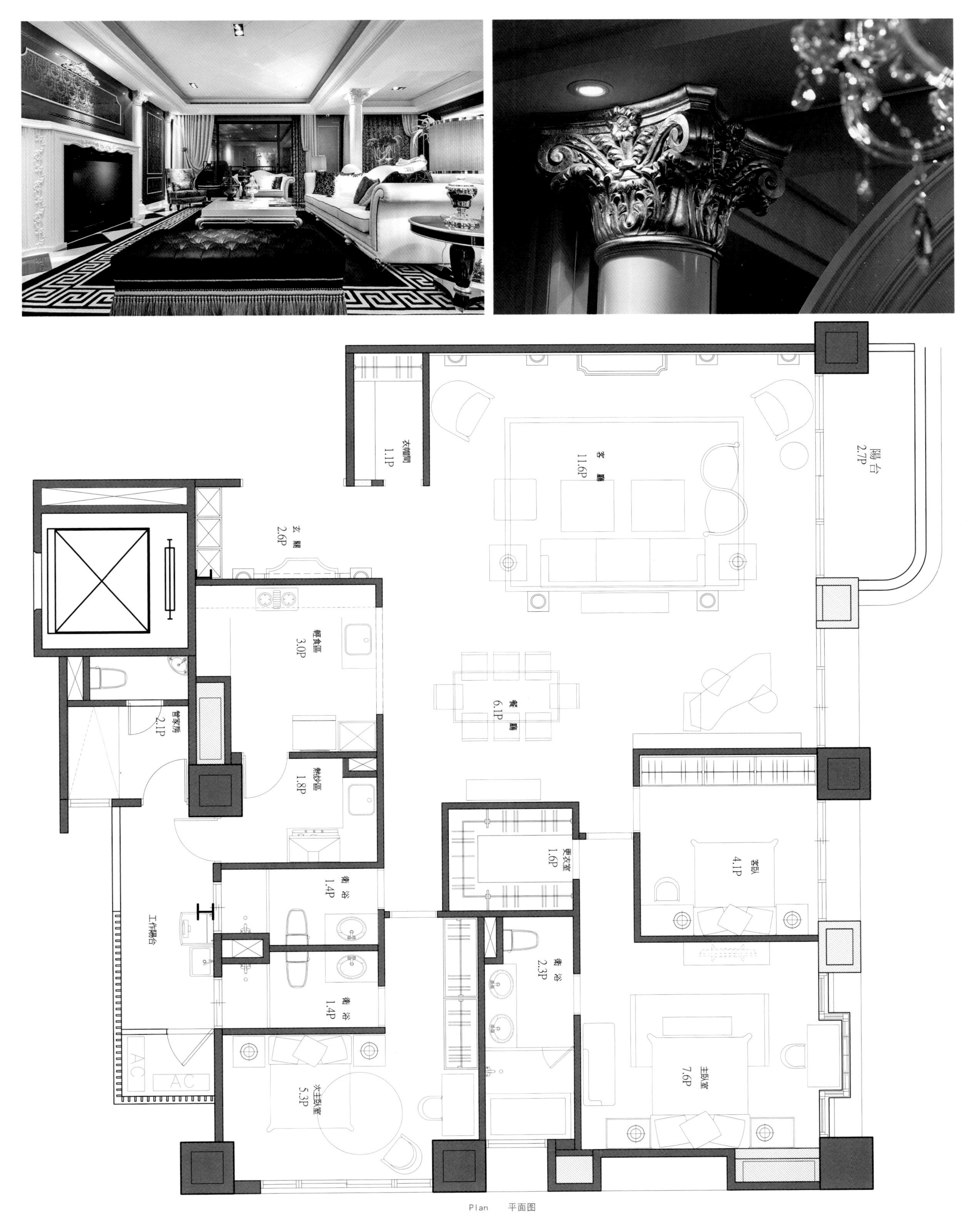

衣帽間
1.1P
客廳
11.6P
陽台
2.7P
玄關
2.6P
輕食區
3.0P
餐廳
6.1P
管家房
2.1P
熱炒區
1.8P
更衣室
1.6P
客臥
4.1P
衛浴
1.4P
衛浴
1.4P
衛浴
2.3P
工作陽台
AC
AC
次主臥室
5.3P
主臥室
7.6P

Plan 平面图

赏析本案，如同观赏一
以浮华人生为主题的超
实歌舞剧，提供观者突
框架的想象力、混合梦
与现实的虚幻效果，以
强烈反差形成的戏剧张
藉由线条、图腾、装饰
家具层层开展，传达空
的丰富动感，让每一位
访者随着空间铺陈而舞
其中。

巴洛克风格的特征是华
力量、富足，服膺着十
世纪的欧洲，体现向外
张、追求财富的时代氛围
一方面发展科学，同时
因为不断征战而动荡，
巴洛克风格喜用繁复、
丽的流动线条表达强烈
情，玄武设计掌握其中
术精神，以黑、灰、白
色彩基调，加上少量金
银勾边与装饰，辅以亮
材质、水晶、玻璃产生
光影，用视觉动静的极
反差，激荡出新奇前卫
巴洛克美学作品。

YUANXIONG · CITY SAMPLE HOUSE

远雄 · 新都样品屋

- Designer: Wang Jishou
- Location: Taipei
- Area: 264 m²

- 设计师：王基守
- 项目地点：台北市
- 项目面积：264 m²

The case, take the essence of Modern Baroque's spirit, discard it's red tape, in this changeable and gorgeous style,the project's colors are no longer showy to pursue the magnificence, but only to reflected in the general details with a modest way.

Designers give a blueprint life with beautiful and elegant for the future master and mistress when they design this case. After choosing the Neoclassicism for the base, simplifying the complicated classic elements and blending the British-style serious atmosphere in this design, then resigner creating a elegant and graceful Neoclassic legend.

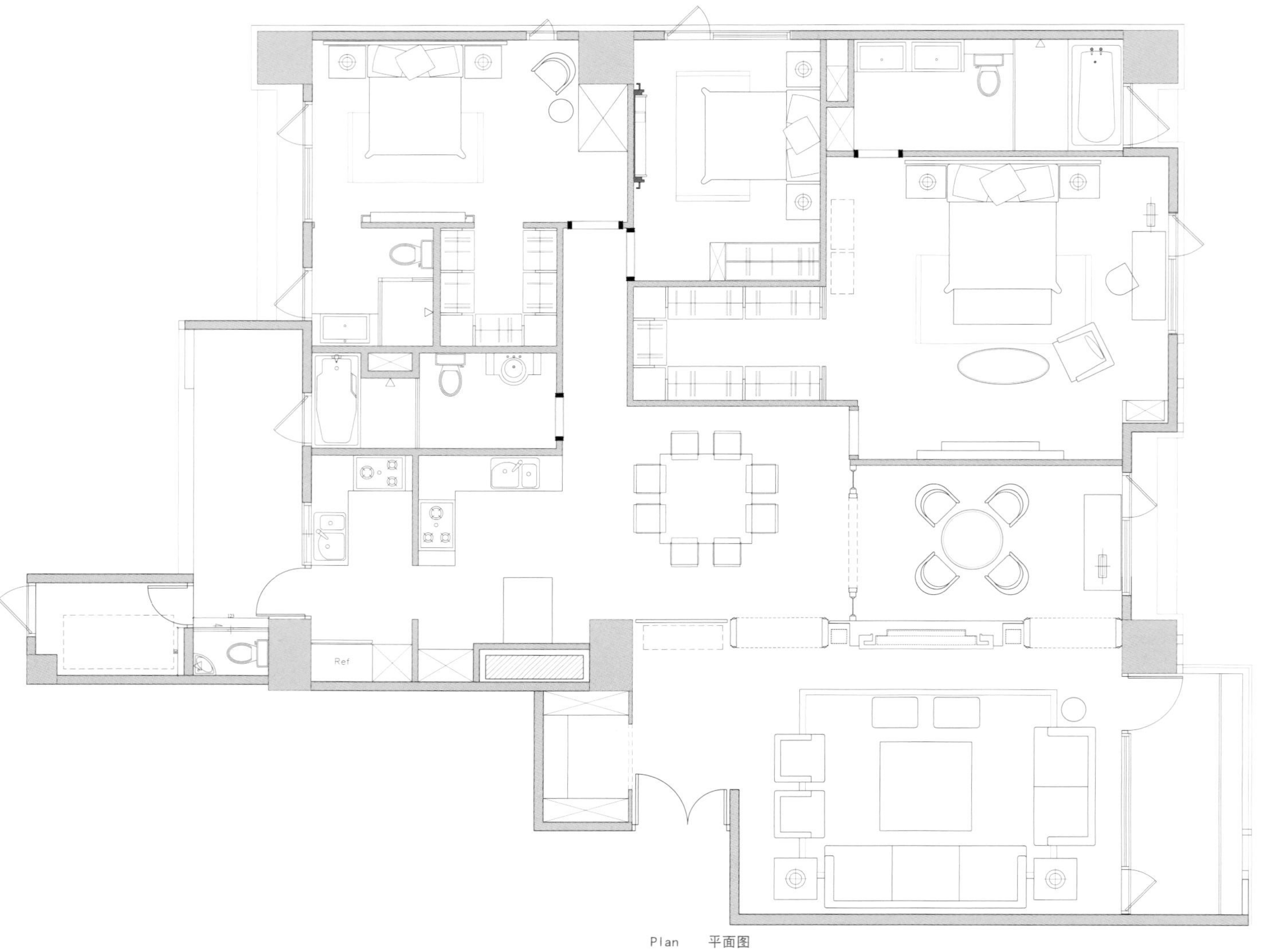

Plan　平面图

本案将巴洛克风格取其精华，去其繁复夸饰，在华丽富有变化的风格中，用色不再夸张，描金也只在细节中含蓄表露。

在构思本案之时，设计师们就为此空间的未来男女主人，勾画出了一幅优雅生活的美丽蓝图。选择新古典为主轴，将繁复的古典语汇简化，并融入英式的庄重气息，进而在这户庄园般的宅邸中，揉合出简约雅致的生活情怀。

HARBIN REAL ESTATE

哈尔滨楼盘项目

- Design Company: Beijing Fashion Imperssion Decoration co.ltd
- Location: Harbin
- Area: 52 m²

- 设计公司：北京风尚印象装饰有限责任公司
- 项目地点：哈尔滨
- 项目面积：52 m²

The residence is tailored by the new-rich born in 1980s, whose remarkable tolerance to novelty and life in the fast lane can be attributed to their generation's shared background, leading to success in both career and social networks.

The project's client virtualizes his vocation as a mass media professionals, who can get the first-hand information owning to his sensitive to fashion and comprehension to new things. Bright and bold colors cater to potential householders' unique personality and fashion appreciation: black and white allow the furniture an intense contrast, while red and champagne gold offer additional embellishment, both resulting in an unrestrained and luxurious atmosphere.

LA SOIREE
OPELLI Collection
2008-2018 Catalogue
FURNITURE HISTORY
2008-2018 Catalogue
2008-2018 Catalogue

Plan 平面图

本案设计对象定位为 80 后单身新贵，由于所处的年代背景，所以接触新鲜事物能力较快，对于快节奏的生活状态适应能力也比较强，属于事业成功的 BB 一族，有一定范围的朋友圈子。

业主虚设为传媒职业，对时尚的感触、对新鲜事物的领悟，都能够及时地掌握第一手资料。快速的生活节奏，男主人用丰富的业余生活调解自己的工作状态。男主人个性鲜明、时尚，在色彩上选用鲜明的色系，突出主人的性格特点。黑白调的家具，对比强烈，红色、香槟金的点缀突出空间气质，在空间整体感上体现出张扬、奢华的氛围。

NEW BAROQUE

新巴洛克风格

- Designer: Luo Yiming (Hanmo Design Studio, Chongqing)
- Location: Chongqing
- Area: 208 m²

- 设计师：罗一鸣（重庆翰墨设计师事务所）
- 项目地点：重庆
- 项目面积：208 m²

The Neo-Baroque residence is a gift loaded with motherly love from the client to her son who is receiving education overseas. A fashionable middle-aged lady, the client enjoys a strong attachment to prestige brands, reserves a dainty taste for life and cherishes vogue and luxury as the ideal design motif. Therefore, the residence becomes ARTDECO oriented where modern concepts and materials rejuvenate European features and contribute to dazzling visual impact.

The project demonstrates apt control and restoration skills, whether in conception or in resulting impression. Pictures fail when conveying the residence's charm that its unique details can only be appreciated on the scene.

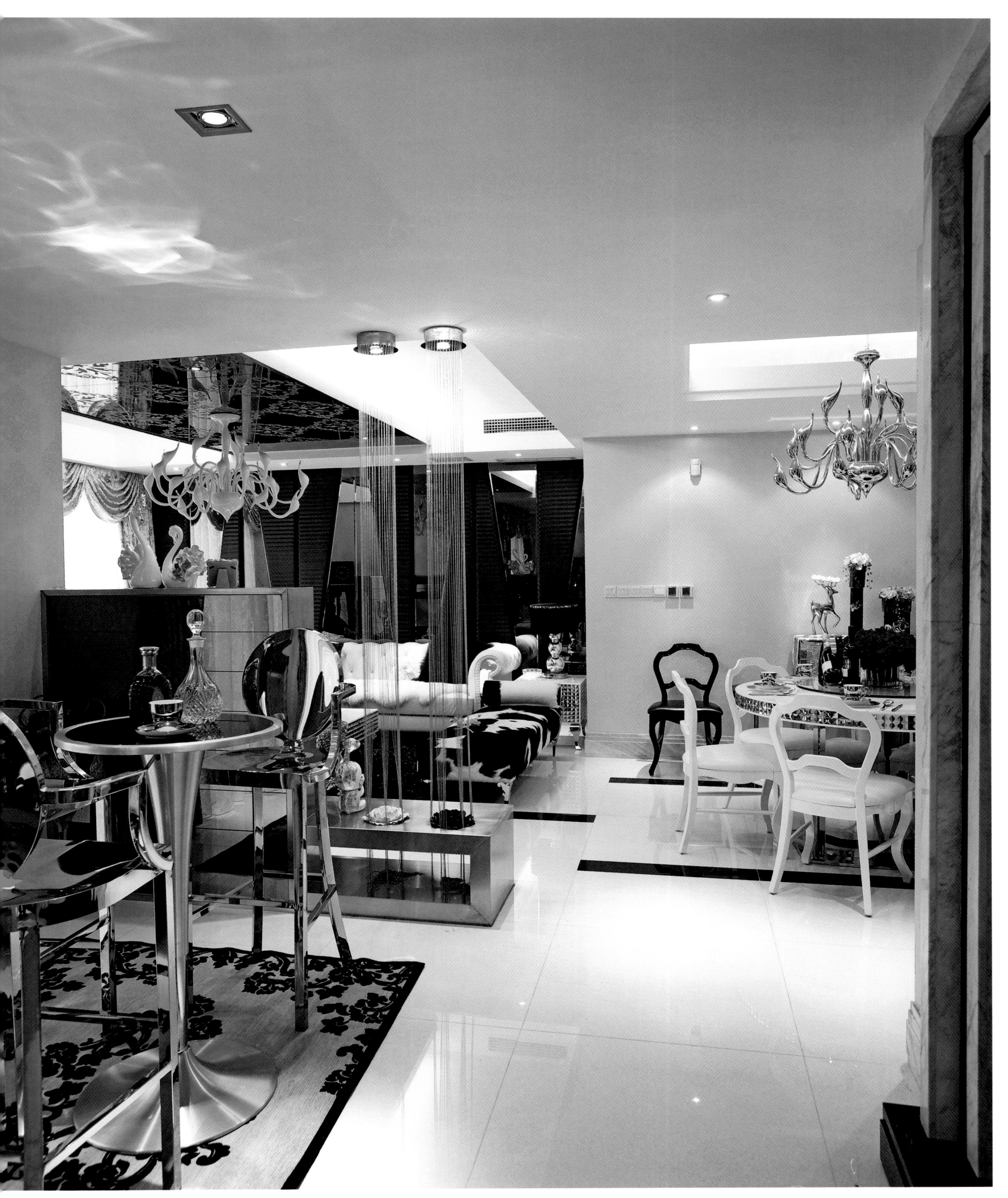

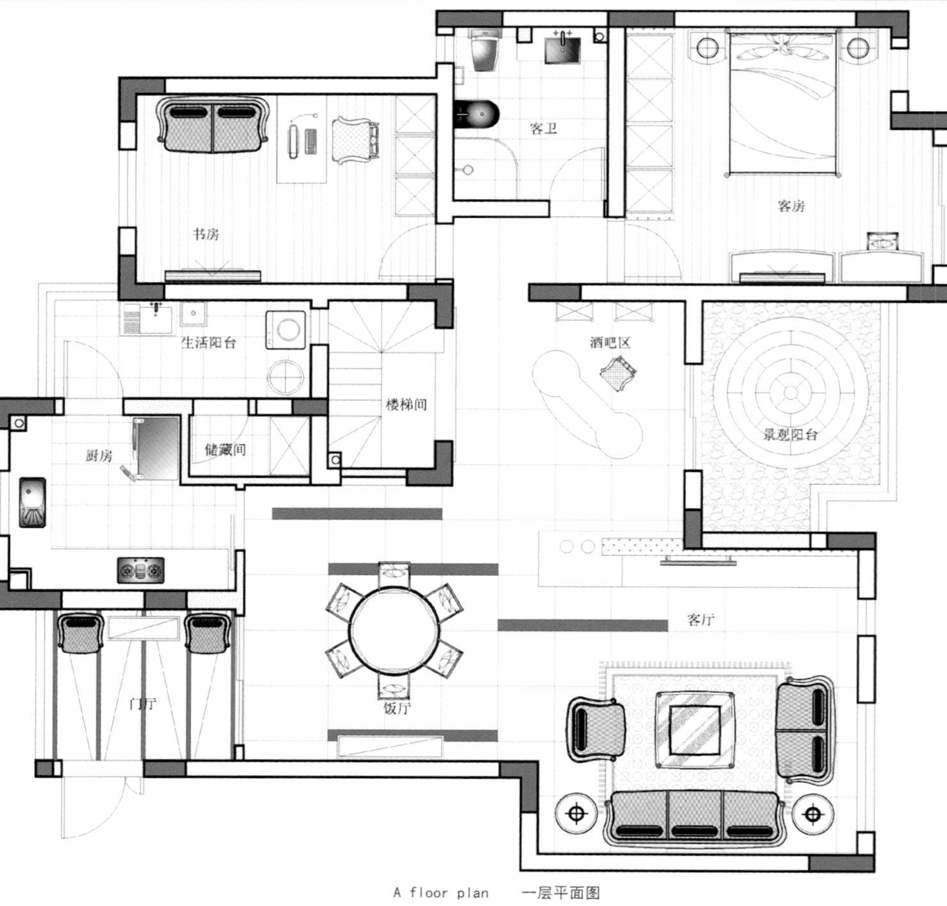

A floor plan　一层平面图

案为新巴洛克风格，业主中年时尚女性，对大牌奢品情有独钟，对生活品质求很高。此套住宅是她为国外留学的儿子购买并装的。时尚而要带一些奢华风格是客户最想要的效果。此我把这套房子风格定位新巴洛克风格。把欧式风用现代的设计手法和材质表达出来，让这种风格上冲突带来令人炫目的视觉果。

设计元素提取到最后效果施，应该说在私宅设计中控度及还原度算比较高的。我们所呈现的只有图片分，其中施工过程的细节，用功能上的很多亮点只有现场才能感受到。

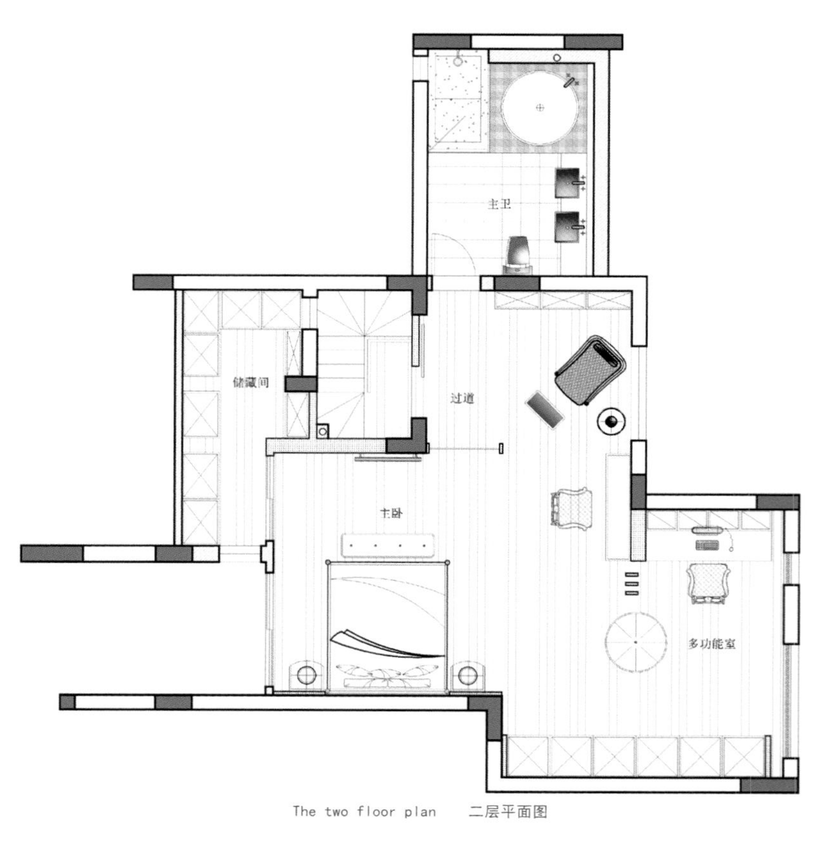

The two floor plan　二层平面图

ELIXIR OF LOVE SONG CONTINUED

花好月圆曲续

- Designer: Liu Weijun (PINKI, Shenzhen; Liu & Associates (IARI) Interior Design Co., Ltd.)
- Location: Xi'an, China
- Area: 250 m²

- 设计师：刘卫军（深圳市品伊创意设计机构&美国IARI刘卫军设计师事务所）
- 项目地点：中国 西安
- 项目面积：250 m²

The upper level flat of the fourplex, situated in Yango Garden of Shanglinfu, Xi'an and dominated by traditional European interior design, is a perfect place for relaxing vacations. The residence reserves exotic nostalgia for cultured householders as well as an atmosphere of happiness, devotion and dreams consistent with family ideals.

The atmosphere, inspired by European countryside style, embraces the residents like a sweet and satisfactory melody saturated with peacefulness and pleasure with a vivid reflection of a tranquil country life.

The space, highlighting the privacy and diversity within a villa, is delicately arranged for family life.

The designer honors simplicity, inartificiality and comfort as his decorating priority. Fine floor tiles are applied in large proportions. Well-chosen palette furthers the pure and natural theme, for natural wood grain, pale green wallpapers, blue medallion patterns and checked or floral fabrics embellish the residence. The fragrance of country life pervades in the gorgeous and saturated colors, floral fabrics, wrought iron items, ceramics and exuberant greenery, contributing to a residence that is a silent tune, praising a life sweet as blossoms and perfect as the full moon.

本案位于西安阳光城上林赋苑，叠拼上户户型，建筑风格以欧式为主，室内风格以异国风情休闲、度假理念为主。阳光、热情、有梦想是全家人的生活指向，异国故乡的风情更是主人的生活品位指向。

在环境风格上，作者以欧式乡村风格为基调，取花好月圆曲作为空间意向，旨在营造出空间的闲适、惬意，突出乡村环境的恬淡与美好。

在空间布局上，作者以家庭生活为导向，突出别墅家庭生活的私密性和多元化

在设计选材上，作者以朴质，自然和舒适为最高原则。运用了大面积的砖，自然的木纹，绿色的墙纸，蓝色的拼花马赛克，方格和小碎花的布艺等。在色彩的选择上自然清新，色彩饱和艳丽，很好地融合了乡村田园的气息。加入一些小碎花、铁艺、陶瓷制品和随处可见的绿色植物都体现着乡村风格的自然和惬意，很好地突显了“花好月圆曲”的空间意向。

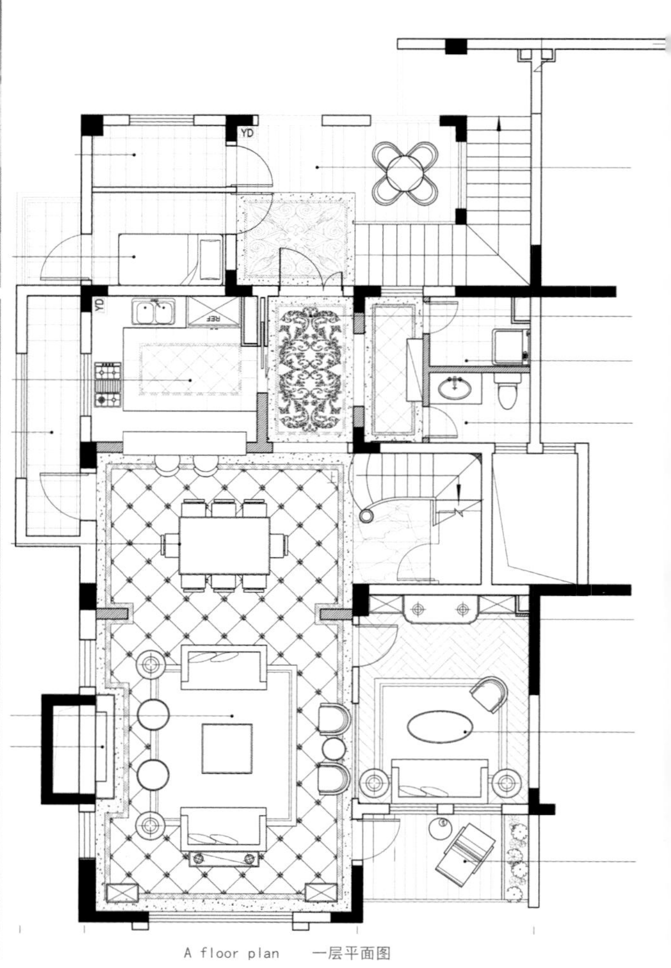

A floor plan　一层平面图

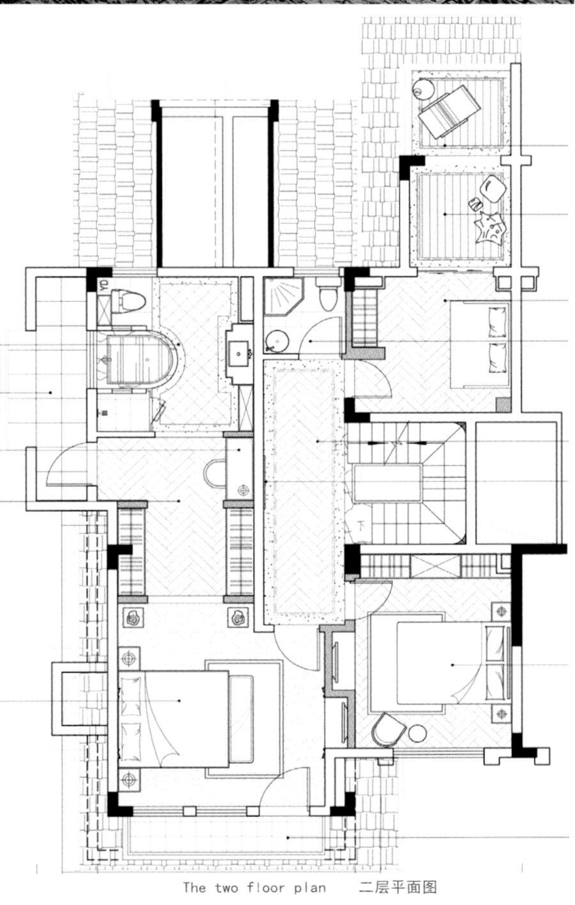

The two floor plan　二层平面图

THE RIPPLING LILY POND

尚漾菁致

- Design Company: PINKI (a design studio); weijun Liu & Associates (IARI) Interior Design Co., LTD.
- Location: Chongqing, China
- Area: 138 m²

- 设计公司：PINKI（品伊）创意机构＆美国IARI刘卫军设计师事务所
- 项目地点：中国 重庆
- 项目面积：138 m²

Responding attentively to the client's taste, fashion, elegance and inspiration are finalized as the design motif. The residence, not overly generous in space, is visually enlarged by dominating light shades. Embellished with modern conciseness, the project qualifies as a haven in the bustling city. Simple partitions retain spatial intactness, while optimized entrance and interior arrangements grant unobstructed movements. The design motif is fully expessed by a quote from the designer: "Vogue's simplicity is beyond expectation."

Varied ceiling materials, providing an obvious difference in spatial styles, ingeniously divide the otherwise integrated dining and drawing room. Entirely covered with a mirror, the dining room ceiling visually heightens the place, while fashionable modern elements and a classic European crystal chandelier are uniquely impressive. Volakas marble is boldly applied in huge proportions, exemplifying the air of modern petite bourgeoisie while increasing the coziness and loveliness of the residence.

We did not grew older we more clearly We did not
grew older we more clearly
FATUATIO

LIFESTYLE
生活

We did not grew older we
more clearly We did not
grew older we more clearly
We did not grew older
clearly We did not
more clearly grew older
FATUATIO

DYNASTY

CONCEPT HALL
NEW YORK CITY
NEW YORK CITY-1938

Martinique

Vin Blanc

时尚、优雅、灵感是本案定下的基本格调。在了解了客户的内心想法后，设计师给出了贴切的回应。本案户型不算大，整体色调考虑以浅色调为主。设计师以简洁现代的设计思想展开，致力将这里打造为繁华都市里的神奇避风港。设计师对整个户型保留完整的空间划分，注重对入户及室内空间的流线优化。“我的目标就是要让大家发现原来家可以如此时尚，而一切又如此简单。”设计师说。

餐厅与客厅在同一空间的情况下，设计师运用天花材质的区分，营造不同空间的感受。餐厅天花的镜面设计，将空间在感官视觉上拉高。时尚现代的造型，搭配欧式古典元素的水晶吊灯使空间更具气质。设计师以夸张大胆的设计手法，将公共空间大面积使用爵士白大理石的同时也将现代都市新贵的小资情结融入其中，使空间更为温馨舒适。

Plan 平面图

ROMANTIC WALTZ

华尔兹恋曲

- Design Company: PINKI & Zhi Ben Jia Soft Decoration Institution
- Location: Chongqing
- Area: 300 m²

- 设计公司：PINKI（品伊）创意机构＆知本家陈设艺术机构
- 项目地点：重庆
- 项目面积：300 m²

A romantic and echoing waltz inspires the design motif. Simple but not prosaic, classic but not stale, the elegant high-end residence is adorned with marquetry down to its last corner. Chaste pale blue, light purple and volakas marble implies feminine tenderness, while nostalgia claret, dark coffee and black represent chevalier virtues. The Chinese kitchen and dining room with rhombus floor tiles are joined by a spacious western kitchen, offering cuisine native to the kitchens' origins. The unique master bedroom, or rather a pinnacle of romance, is embellished with laced valance, jet-black crystal chandelier, garnet night tables and a red velvet bed stool.

Come waltzing to the melody of dream, where stiletto heels and shinny shoes imprint footsteps on the journey of life, where gentle touches and accelerated heartbeats proclaim a romance of destiny. Come waltzing till the end of life, lest the romance forever fades.

本案的空间设计正是如同一场浪漫回旋的华尔兹。整体设计简约而不简单，去掉腐朽、保留经典，从整体到局部，精雕细琢，镶花刻金，典雅而高贵。客厅里，优雅的淡蓝、浅紫、爵士白，如同女性的浪漫情愫；怀旧的深啡、暗红、高更黑，如同男性的绅士情怀。菱形花砖的中式厨房与餐厅里摆放着美酒、舒展大方的西厨让主人可以尽享中、西餐饮。主卧床幔帘头花边、黑玫瑰色的水晶宫灯、暗红床头柜、银绒红床尾凳的质感展现出独特的气质，言说浪漫，亦不过如此。

他们在旋舞中梦回几转，在身体的轻触，旋转的脚步中重读了前世的絮语，男人的皮鞋，女人细跟高跟鞋，逾越了一场命运的轮回。如果可以，这场华尔兹便成为了永恒，但却是你回眸转瞬即逝，再回首便已是前世今生。

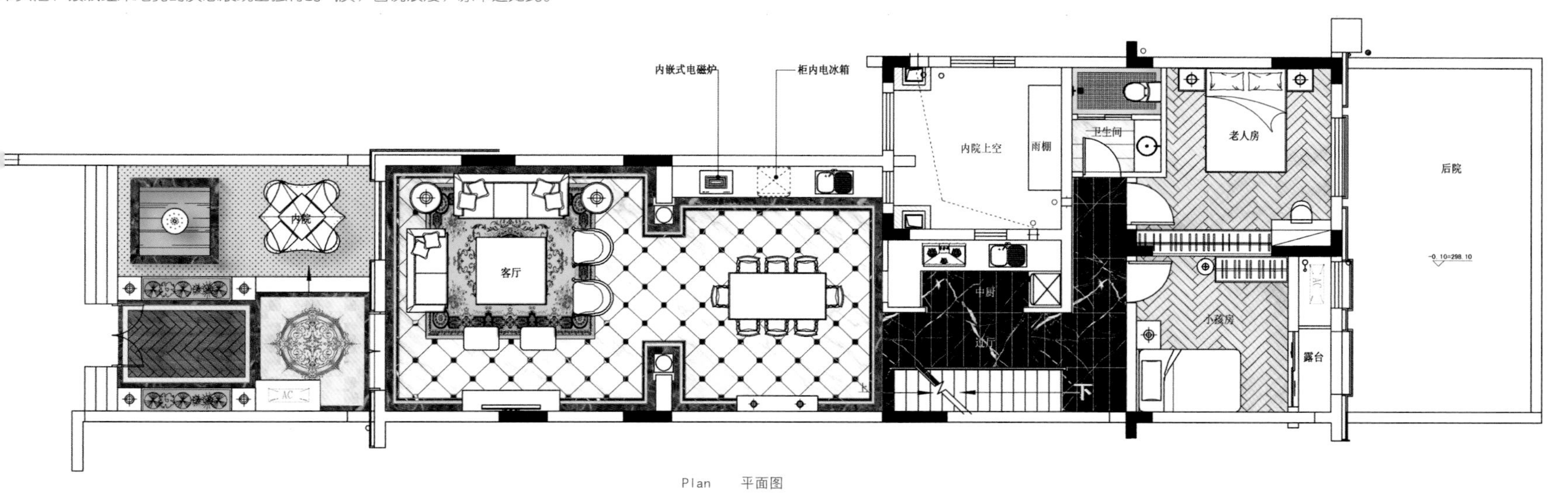

Plan　平面图

图书在版编目(CIP)数据

再定义奢华宅邸/深圳市海阅通文化传播有限公司 编著. —武汉：华中科技大学出版社, 2013.2
ISBN 978-7-5609-8681-4

Ⅰ.①再… Ⅱ.①深… Ⅲ.①住宅—室内装饰设计 Ⅳ.①TU241

中国版本图书馆CIP数据核字(2013)第030435号

再定义奢华宅邸

深圳市海阅通文化传播有限公司 编著

出版发行：华中科技大学出版社（中国·武汉）
地　　址：武汉市武昌珞喻路1037号（邮编：430074）
出 版 人：阮海洪

责任编辑：赵慧蕊　　责任监印：张贵君
责任校对：张雪姣　　装帧设计：陈秋娣

印　　刷：中华商务联合印刷（广东）有限公司
开　　本：965 mm×1270 mm　1/16
印　　张：21
字　　数：302千字
版　　次：2013年5月第1版 第1次印刷
定　　价：358.00元　（USD 71.99）

投稿热线：(027)87545012　design_book_wh01@hustp.com
本书若有印装质量问题，请向出版社营销中心调换
全国免费服务热线：400-6679-118 竭诚为您服务